FROM MOLECULES TO MINDS

Challenges for the 21st Century

WORKSHOP SUM

Matthew Hougan and Bruce Altevc   *pporteurs*

Forum on Neuroscience and Nervous System Disorders

Board on Health Sciences Policy

INSTITUTE OF MEDICINE
OF THE NATIONAL ACADEMIES

THE NATIONAL ACADEMIES PRESS
Washington, D.C.
www.nap.edu

THE NATIONAL ACADEMIES PRESS • 500 Fifth Street, N.W. • Washington, DC 20001

This project was supported by contracts between the National Academy of Sciences and the Alzheimer's Association; AstraZeneca Pharmaceuticals, Inc.; CeNeRx Biopharma; the Department of Health and Human Services' National Institutes of Health (NIH, Contract No. N01-OD-4-213) through the National Eye Institute, the National Institute of Mental Health, the National Institute of Neurological Disorders and Stroke, the National Institute on Aging, the National Institute on Alcohol Abuse and Alcoholism, the National Institute on Drug Abuse, and the NIH Blueprint for Neuroscience Research; Eli Lilly and Company; GE Healthcare, Inc.; GlaxoSmithKline, Inc.; Johnson & Johnson Pharmaceutical Research and Development, Inc.; Merck Research Laboratories, Inc.; the National Multiple Sclerosis Society; the National Science Foundation (Contract No. OIA-0753701); the Society for Neuroscience; and Wyeth Research, Inc. The views presented in this publication are those of the editors and attributing authors and do not necessarily reflect the view of the organizations or agencies that provided support for this project.

International Standard Book Number-13: 978-0-309-12092-0
International Standard Book Number-10: 0-309-12092-6

Additional copies of this report are available from The National Academies Press, 500 Fifth Street, N.W., Lockbox 285, Washington, DC 20055; (800) 624-6242 or (202) 334-3313 (in the Washington metropolitan area); Internet, http://www.nap.edu.

For more information about the Institute of Medicine, visit the IOM home page at: **www.iom.edu.**

Suggested citation: IOM (Institute of Medicine). 2008. From molecules to minds: Challenges for the 21st century: Workshop summary. Washington, DC: The National Academies Press.

"Knowing is not enough; we must apply.
Willing is not enough; we must do."

—Goethe

INSTITUTE OF MEDICINE
OF THE NATIONAL ACADEMIES

Advising the Nation. Improving Health.

THE NATIONAL ACADEMIES
Advisers to the Nation on Science, Engineering, and Medicine

The **National Academy of Sciences** is a private, nonprofit, self-perpetuating society of distinguished scholars engaged in scientific and engineering research, dedicated to the furtherance of science and technology and to their use for the general welfare. Upon the authority of the charter granted to it by the Congress in 1863, the Academy has a mandate that requires it to advise the federal government on scientific and technical matters. Dr. Ralph J. Cicerone is president of the National Academy of Sciences.

The **National Academy of Engineering** was established in 1964, under the charter of the National Academy of Sciences, as a parallel organization of outstanding engineers. It is autonomous in its administration and in the selection of its members, sharing with the National Academy of Sciences the responsibility for advising the federal government. The National Academy of Engineering also sponsors engineering programs aimed at meeting national needs, encourages education and research, and recognizes the superior achievements of engineers. Dr. Charles M. Vest is president of the National Academy of Engineering.

The **Institute of Medicine** was established in 1970 by the National Academy of Sciences to secure the services of eminent members of appropriate professions in the examination of policy matters pertaining to the health of the public. The Institute acts under the responsibility given to the National Academy of Sciences by its congressional charter to be an adviser to the federal government and, upon its own initiative, to identify issues of medical care, research, and education. Dr. Harvey V. Fineberg is president of the Institute of Medicine.

The **National Research Council** was organized by the National Academy of Sciences in 1916 to associate the broad community of science and technology with the Academy's purposes of furthering knowledge and advising the federal government. Functioning in accordance with general policies determined by the Academy, the Council has become the principal operating agency of both the National Academy of Sciences and the National Academy of Engineering in providing services to the government, the public, and the scientific and engineering communities. The Council is administered jointly by both Academies and the Institute of Medicine. Dr. Ralph J. Cicerone and Dr. Charles M. Vest are chair and vice chair, respectively, of the National Research Council.

www.national-academies.org

WORKSHOP ON GRAND CHALLENGES IN NEUROSCIENCE PLANNING COMMITTEE[*]

ALAN LESHNER (*Chair*), American Association for the Advancement of Science, Washington, DC
ALAN BREIER, University of Indiana, Indianapolis
DAVID COHEN, Columbia University, Society for Neuroscience representative, New York
RICHARD HODES, National Institute on Aging, Bethesda, MD
STEVEN HYMAN, Harvard University, Cambridge, MA
JUDY ILLES, University of British Columbia, Vancouver, Canada
THOMAS INSEL, National Institute of Mental Health, Bethesda, MD
STORY LANDIS, National Institute of Neurological Disorders and Stroke, Bethesda, MD
TING-KAI LI, National Institute on Alcohol Abuse and Alcoholism, Bethesda, MD
MICHAEL OBERDORFER, National Institutes on Health Neuroscience Blueprint, Bethesda, MD
KATHIE OLSEN, National Science Foundation, Arlington, VA
WILLIAM POTTER, Merck Research Laboratories, Inc., North Wales, PA
ROBERT RICHARDSON, Cornell University, Ithaca, NY
PAUL SIEVING, National Eye Institute, Bethesda, MD
RAE SILVER, Columbia University, New York
ROY TWYMAN, Johnson & Johnson Pharmaceutical Research and Development, Inc., Titusville, NJ
NORA VOLKOW, National Institute on Drug Abuse, Bethesda, MD

IOM Staff

BRUCE ALTEVOGT, Forum Director
SARAH HANSON, Senior Program Associate
LORA TAYLOR, Senior Project Assistant
DIONNA ALI, Anderson Intern

---

[*]IOM planning committees are solely responsible for organizing the workshop, identifying topics, and choosing speakers. The responsibility for the published workshop summary rests with the workshop rapporteurs and the institution.

FORUM ON NEUROSCIENCE
AND NERVOUS SYSTEM DISORDERS[*]

ALAN LESHNER (*Chair*), American Association for the Advancement of Science, Washington, DC
HUDA AKIL, University of Michigan–Ann Arbor
MARC BARLOW, GE Healthcare, Inc., Buck, United Kingdom
DANIEL BURCH, CeNeRx Biopharma, Research Triangle Park, NC
DENNIS CHOI, Emory University, Atlanta, GA
TIMOTHY COETZEE, National Multiple Sclerosis Society, New York
DAVID COHEN, Columbia University, Society for Neuroscience representative, New York
RICHARD FRANK, GE Healthcare, Inc., Princeton, NJ
RICHARD HODES, National Institute on Aging, Bethesda, MD
STEVEN HYMAN, Harvard University, Cambridge, MA
JUDY ILLES, University of British Columbia, Vancouver, Canada
THOMAS INSEL, National Institute of Mental Health, Bethesda, MD
STORY LANDIS, National Institute of Neurological Disorders and Stroke, Bethesda, MD
TING-KAI LI, National Institute on Alcohol Abuse and Alcoholism, Bethesda, MD
HUSSEINI MANJI, Johnson & Johnson Pharmaceutical Research and Development, Inc., Titusville, NJ (since September 2008)
MICHAEL OBERDORFER, National Institutes on Health Neuroscience Blueprint, Bethesda, MD
KATHIE OLSEN, National Science Foundation, Arlington, VA
ATUL PANDE, GlaxoSmithKline, Inc., Research Triangle Park, NC
MENELAS PANGALOS, Wyeth Research, Inc., Princeton, NJ
STEVEN PAUL, Eli Lilly and Company, Indianapolis, IN
WILLIAM POTTER, Merck Research Laboratories, Inc., North Wales, PA
SCOTT REINES, Foundation for the NIH, Bethesda, MD
PAUL SIEVING, National Eye Institute, Bethesda, MD
RAE SILVER, Columbia University, New York
WILLIAM THIES, Alzheimer's Association, Chicago, IL

---

[*]IOM forums and roundtables do not issue, review, or approve individual documents. The responsibility for the published workshop summary rests with the workshop rapporteurs and the institution.

ROY TWYMAN, Johnson & Johnson Pharmaceutical Research and
    Development, Inc., Titusville, NJ (until September 2008)
NORA VOLKOW, National Institute on Drug Abuse, Bethesda, MD
FRANK YOCCA, AstraZeneca Pharmaceuticals, Wilmington, DE
CHRISTIAN ZIMMERMAN, Neuroscience Associates, Boise, ID

IOM Staff

BRUCE ALTEVOGT, Forum Director
SARAH HANSON, Senior Program Associate
LORA TAYLOR, Senior Project Assistant
DIONNA ALI, Anderson Intern

IOM Staff

ANDREW M. POPE, Director
AMY HAAS, Board Assistant
DONNA RANDALL, Financial Associate

Independent Report Reviewers

This report has been reviewed in draft form by individuals chosen for their diverse perspectives and technical expertise, in accordance with procedures approved by the National Research Council's Report Review Committee. The purpose of this independent review is to provide candid and critical comments that will assist the institution in making its published report as sound as possible and to ensure that the report meets institutional standards for objectivity, evidence, and responsiveness to the study charge. The review comments and draft manuscript remain confidential to protect the integrity of the deliberative process. We wish to thank the following individuals for their review of this report:

William Bialek, Department of Physics, Princeton University
Fred H. Gage, The Salk Institute for Biological Studies
David J. Kupfer, Western Psychiatric Institute and Clinic, University of Pittsburgh School of Medicine
Tom M. Mitchell, Machine Learning Department, Carnegie Mellon University
Robert C. Richardson, Cornell University

Although the reviewers listed above have provided many constructive comments and suggestions, they did not see the final draft of the report before its release. The review of this report was overseen by **Dr. Sid Gilman,** Michigan Alzheimer's Disease Research Center, Department of Neurology, University of Michigan. Appointed by the Institute of Medicine, he was responsible for making certain that an independent examination of this report was carried out in accordance with institutional procedures and that all review comments were carefully considered. Responsibility for the final content of this report rests entirely with the authoring committee and the institution.

Contents

Workshop Summary

INTRODUCTION[1]

The goals of medicine are to "wrest from nature the secrets which have perplexed philosophers of all ages. . . ."

—*Sir William Osler, 1849–1919*

On June 25, 2008, more than 70 of the leading neuroscientists in the world gathered at the National Academy of Sciences building in Washington, DC, for a workshop hosted by the Institute of Medicine's (IOM's) Forum on Neuroscience and Nervous System Disorders titled, "From Molecules to Mind: Challenges for the 21st Century." Their goals were significant: Each participant was asked to identify one or two "Grand Challenges" that could galvanize both the scientific community and the public around the possibilities for neuroscience in the 21st century.

This idea of identifying Grand Challenges has a strong history in science. For example, as Kathie Olsen, deputy director of the National Science Foundation, reminded the panelists, the physics community was united in 2003 by the publication of *Connecting Quarks with the Cosmos.* This National Research Council (NRC) committee report identified a handful of fundamental questions about the universe, such as "What powered the big bang?" and "What is dark matter?" (NRC, 2003). More

---

[1]The planning committee's role was limited to planning the workshop, and the workshop summary has been prepared by the workshop rapporteurs as a factual summary of what occurred at the workshop.

1

recently the National Academy of Engineering developed a set 14 Grand Challenges for engineering in the 21st century (NRC, 2008).[2]

In each case, a common purpose—combined with new funding, new technologies, new ideas, and an influx of new scientists—drove researchers to tackle problems that seemed impossible just a few years earlier.

Neuroscience has made phenomenal advances over the past 50 years and the pace of discovery continues to accelerate. Some of that progress has resulted from the simultaneous appearance of new technologies, like those of molecular biology, neuroimaging, and computer and information science. The progress of the past in combination with these new tools and techniques has positioned neuroscience on the cusp of even greater transformational progress in our understanding of the brain and how its activities result in mental activity.

Recognizing that neuroscience is not, of course, really a single field is important. Rather, it is a multidisciplinary enterprise including diverse fields of biology, psychology, neurology, chemistry, mathematics, physics, engineering, computer science, and more. If scientists within neuroscience and related disciplines could unite around a small set of goals, the opportunity for advancing our understanding of brain and mental function would be huge.

Exploring that potential set of common goals, or Grand Challenges, was one of the major goals of the workshop.

What Can We Achieve

For a Grand Challenges exercise to work, it must ask questions that are both big *and* answerable. The questions must fire the soul and stir the spirit, but also be approachable in a scientifically rigorous manner, explained Alan Leshner, chief executive officer of the American Association for the Advancement of Science and chair of the Forum on Neuroscience.

For neuroscience, the first part is easy. Neuroscience is aimed at one of the most fundamental questions of all: How does our physical body

---

[2]*Connecting Quarks with the Cosmos: Eleven Science Questions for the New Century* was authored by the NRC's Committee on the Physics of the Universe. *Grand Challenges for Engineering* was authored by the NRC's Committee on Grand Challenges for Engineering. Neither was a workshop summary, but rather included specific findings and recommendations of the respective committee.

give rise to a person who can think, love, learn, and dream? Hippocrates identified the brain as the seat of human experience in 400 B.C., and we have been trying to figure out how it works ever since.

However, as was demonstrated throughout the workshop and as will be highlighted throughout this workshop summary, neuroscience has advanced to the point where answering those questions in a rigorous manner is truly possible, commented Leshner.

By the end of the workshop there was a sense of momentum and of new frontiers opening up, remarked Leshner. The brain is one of the most complicated and exquisite objects on earth. According to Colin Blakemore, a leading British neuroscientist from Oxford University and the former chief executive of the British Medical Research Council, "There are more neurons in the brain than there are stars in the galaxy."

Who among us has not wondered how it all works; how the lump of our physical brain gives rise to someone who can want, and love, and read poetry?

About This Workshop

The Neuroscience Challenges for the 21st Century workshop was hosted by the IOM's Forum on Neuroscience and Nervous Systems Disorders, which is a convening activity at the IOM dedicated to furthering our understanding of the brain and nervous systems, disorders in their structure and function, and effective clinical prevention and treatment strategies. The Forum brings together experts from private-sector sponsors of biomedical and clinical research, federal agencies sponsoring and regulating biomedical and clinical research, foundations, the academic community, and consumers to talk about issues of mutual interest and concern.

The goals of a forum, and this workshop, are not to provide specific recommendations or arrive at consensus conclusions; rather, a forum seeks to highlight important issues and articulate the challenges facing a particular scientific field. Organized by an independently appointed planning committee, the workshop was organized so that representatives of all corners of the neuroscience world could provide updates on the latest advances in the field, and then discuss how they related to the concept of Grand Challenges. Throughout the day, participants learned how advances in imaging technology, computer science, molecular biology, biochemistry, and neuroscience in general had made it possible for us to imagine understanding how the brain works at a fundamental level—something that was not possible just 2 or 3 years ago.[3] In addition, each participant was invited to present his/her impression of what one or two Grand Challenges would be for the neurosciences. As a result, throughout this document key insights are attributed to at least one participant. When multiple parties were involved in fashioning or honing a single idea or insight, the author has endeavored to attribute that idea or insight to the key parties involved.

Leshner and Olsen concluded the workshop by synthesizing the day's discussions into three overarching Grand Challenges that emerged during the workshop, which will be used to organize this workshop summary:

- How does the brain work and produce mental activity? How does physical activity in the brain give rise to thought, emotion, and behavior?
- How does the interplay of biology and experience shape our brains and make us who we are today?
- How do we keep our brains healthy? How do we protect, restore, or enhance the functioning of our brains as we age?

In addition, this summary includes a synopsis of topics that emerged during the discussion that do not fall specifically under any one of the three Grand Challenge questions identified here, including some challenges and technical limitations as well as ethical concerns.

GRAND CHALLENGE: HOW DOES THE HUMAN BRAIN WORK AND PRODUCE MENTAL ACTIVITY?

How does the brain work and produce mental activity? How does physical activity in the brain give rise to thought, emotion, and behavior?

We envision our brains taking in data, running those data through some unknown processes, and then somehow telling us how to act, feel, or behave. "What are the algorithmic principles that the brain uses?" Blakemore asked. "Are there some which are nonalgorithmic? How can we approach the modeling of those principles?"

In the deepest sense, we do not know how information is processed, stored, or recalled; how motor commands emerge and become effective;

---

[3]To download presentations or listen to audio archives, please visit http://www.iom.edu/CMS/3740/35684/54555.aspx.

how we experience the sensory world; how we think or feel or empathize. This is because explanations ultimately must be integrated across levels of analysis, including: molecular, cellular, synaptic, circuit, systems, computational, and psychological, and until now the field has not been mature enough to integrate information across all these disciplines.

These are some of the most compelling questions in the world, said Olsen in the opening session of the workshop.

Of course there is another reason—or rather, many millions of reasons—why we do not have a working theory of the brain. As Blakemore pointed out, there are more neurons in the brain than there are stars in the galaxy, and we form more than 1 million new connections among these neurons each day. Simply put, the scope of the challenge is awesome.

Still, the feeling among many at the workshop was that there was hope in meeting this challenge.

The reason? Major technological advances during the past few years are allowing neuroscientists to do the kind of research and tackle the kind of challenges they have always dreamed of, starting, according to many at the workshop, with drawing up the wiring diagram of the human brain.

Mapping the Human Brain

The idea of mapping the human brain is not new. The "father of neuroscience," Santiago Ramon y Cajal, argued at the turn of the 20th century that the brain was made up of neurons woven together in a highly specific way. We have been trying to map this exquisite network since then.

In fact, scientists in other settings have called the wiring diagram a Grand Challenge of neuroscience in and of itself. It appears on the Grand Challenges of the Mind and Brain list for the National Science Foundation (NSF, 2006), on the Grand Challenges list of the National Academy of Engineering (NRC, 2008), and on the wish lists of at least a half-dozen major scientific fields, from genetics to computer science.

If we are interested in how the mind works, then we definitely need to know the physical instantiation of brains and function, remarked Jeffrey Lichtman, professor of molecular and cellular biology, Harvard University. This effort will require some mechanism to obtain the connectional maps that will integrate anatomy, neuronal activity, and function. Until those are available, the field will not be able to move forward to its full potential.

The challenge is similar, in many ways, to mapping the human genome: We might not know exactly what we will learn, but we have a strong belief that we will learn a lot, commented Leshner.

So why has it not happened?

Because neurons are very small and the human brain is exquisitely complex and hard to study. Eve Marder, professor of neuroscience at Brandeis University and president of the Society for Neuroscience, noted that scientists have been working on circuit analysis for nearly 40 years, primarily with smaller organisms, particularly invertebrates, because their simpler neurological systems are more amenable to study and analysis.

The classic approach, in place since the 1960s, has been simple: Define behaviors, identify neurons involved in those behaviors, determine the connectivity between those neurons, and then excite individual neurons to understand their role in influencing behavior. This approach is called "circuit dynamics," and it has been tremendously helpful to understanding how these simple neurological systems work.

But as you move from sponges and anemones to primates and humans, each step of that analytical process becomes infinitely more challenging.

As Marder noted, the impediments, until today, to understanding larger circuits and vertebrate brains include difficulty in identifying neurons, difficulty in perturbing individual classes of neurons in isolation, and difficulty in recording from enough of the neurons at the same time with enough spatial and temporal resolution.

In other words, difficulty arose in every step of the circuit dynamics process.

But the key words in Marder's statement are "until today." If you look at the three things Marder identified as stumbling blocks, major technological breakthroughs over the past few years have solved or are close to solving each one, starting with a new technique born from the lab of Lichtman: "the Brainbow."

Technological Advance: The Brainbow

Mapping the brain is not easy. Neurons and the connections between them are so small and complex that tracing their path through the brain has been nearly impossible.

For more than a century, the best method available to researchers has been the "Golgi stain." Developed in 1873 (and little improved on since), the Golgi method uses a stain of silver chromate salt to trace the path of individual neurons, right down to the axons and dendrites.

The Golgi method works quite well, but comes with two major flaws that limit its use in studying complex connections among neurons in a single network. The first flaw is that the method stains everything the same color—grey—making it very difficult to study multiple neurons at once or to envision how different neurons link together. Second, it is difficult to target specific cells to be stained, that is, neurons that are stained are done so in a largely random pattern.

Over the years, researchers have improved on the Golgi stain. For example, geneticists found ways to "tag" different neurons with genes that naturally produce fluorescent colors, so that the neurons themselves could be made to glow red, blue, or yellow. This advance allowed researchers to study a handful of neurons at once.

The neurosciences have now matured to the point where scientific knowledge and technological advances are converging to bring new capabilities. For example, in 2007 Harvard University researcher Jean Livet, working out of Lichtman's lab, published a paper showing how fluorescent-coding genes from jellyfish and coral could be combined to force different neurons to express hundreds of different colors (Livet et al., 2007). This Brainbow technique relies on three genes—coding for red, blue, and yellow—which are combined in different levels to produce all the different tones. A cell might have three red genes, two blue, and one yellow, for instance.

The result? Researchers can, for the first time, identify and map hundreds of neurons at once, seeing how they wrap and interact with one another, tracing the map of the brain in greater detail than was possible just 1 or 2 years earlier.

Technological Advance: Neuronal "Light Switch"

Marder's second impediment was the challenge of perturbing individual neurons. Even if you can see the connections between the actual cells, if you want to see how one neuron connects to and influences another, and most importantly what impact that has on behavior, you must be able to "excite" those neurons to find out. Over and over and over again.

The classical method uses electrodes to stimulate neurons, but it is neither precise nor particularly sophisticated. Neurons are so small and make so many connections—an individual neuron can make well over 100 separate connections with other neurons—that it is extremely difficult to precisely activate a single neuron, let alone a specific neuronal connection, in an *in-vitro* model system, and even more so in an *in-vivo* vertebrate nervous system.

In 2005, however, researchers in Stanford University and the Max Planck Institute of Biophysics Germany developed a neuronal "light switch" that allows them to turn individual neurons or neuronal connections on or off by exposing them to light (Boyden et al., 2005). The science behind the study is impressive. Researchers discovered a protein from green algae that switches the electrical state of a cell when exposed to blue light. By inserting this gene into rat neurons, researchers were able to gain control over those neurons and consequently their connections, turning them on and off with the flip of a switch. As an added bonus, researchers attached this protein to a gene that glows when exposed to green light, allowing them to both identify and control individual neurons. Therefore, under green light researchers can view the neurons that make the protein, and by switching the light beam to blue, they can excite a neuron and investigate its effects.

The applications and implications of this new technique are many. From a research perspective, being able to turn individual neurons on and off allows advanced study of the function of individual neurons in the brain. From a clinical perspective, the ability to modulate neurons using something as simple and noninvasive as light opens up opportunities for extremely targeted therapies for diseases such as Parkinson's, depression, and more.

Technological Challenge: Spatial and Temporal Resolution

Marder's third challenge—the difficulty in recording from enough of the neurons at the same time with enough spatial and temporal resolution—remains a major challenge for the field. Both imaging and electrode recording capabilities have come a long way in recent years, but multiple researchers expressed the need for more.

Multichannel Microelectrode Recording Arrays

The development of multichannel microelectrode recording arrays allows researchers to accurately measure the activity of multiple neurons at a single time. Advances in photonics, electronic circuitry, and engineering have made it possible for these arrays to be shrunken substantially, dramatically increasing the number of neurons that can be monitored directly through the skin. Moreover, researchers believe the devices can now be implanted in the brain, or else where in the nervous system, suggesting we could measure the output of neurons on an individual level over long periods of time (Kelly et al., 2007).

If we are going to get a real map of the functional wiring diagram of the human, we need to be able to do it noninvasively and on a widespread basis.

"Brain functions are encoded in a distributed network in the brain," said Bin He, professor of biomedical engineering, electrical engineering, and neuroscience, University of Minnesota, so it is important to image brain connectivity and network dynamics not only beyond localized circuits, but throughout the entire network.

Functional Magnetic Resonance Imaging

Functional Magnetic Resonance Imaging (fMRI) allows researchers to noninvasively measure blood flow and blood oxygenation in the brain. Because blood flow and oxygenation are closely linked with brain activity, researchers can see which areas of the brain are active when volunteers (or research animals) are performing an assigned task.

A circuit map that does not correlate back to activity is not extremely valuable. fMRI is one technique used to integrate anatomy back to function, allowing this correlation. Unfortunately, fMRI readings are not perfect. Spatial resolution has only recently advanced to the millimeter level, and unfortunately the measurements are not in real time. There is a delay of about a second between brain activity and associated changes in blood flow and oxygenation that can be detected by the fMRI. However, researchers need to be able to measure activity in a real-time, millisecond-by-millisecond basis and on a much smaller spatial scale. As a result, they are now working on ways to combine fMRI readings with instantaneous feedback loops such as electroencephalography (EEG) and magnetoencephalography (MEG).

"Can we develop a technique which can noninvasively image the neural activity at millimeter spatial resolution and millisecond temporal resolution?" asked He, in a comment echoed by others at the workshop. But even this is resolution is course relative to the size of a neuron—a cubic millimeter of brain cortex contains 10^4 to 10^5 neurons.

Computer Science and Learning Algorithms

Even with all these advances in collecting data, the challenges of mapping the brain remain enormous. The human genome project would not have been possible until the turn of the 21st century, as the genetics field simply did not have the automated techniques or the computer power to tackle the project. The amount of data involved in mapping the structure of the brain is likely to be an order of magnitude greater than was required for mapping the genome, and will require enormous computing capacity. This is where computer science comes in.

One example of using computational methods to link neural activity to psychological states was provided by Tom Mitchell, chair of the Machine Learning Department at Carnegie Mellon University, who described how, through the use of machine learning methods, a person's neural activity and reactions to words or pictures can be decoded via fMRI. Such computer algorithms, which have been adopted by researchers studying brain-wide neural representations, provide a direct link between the biology of neural activity and abstract mental states such as thinking about an object.

In addition, the work of Sebastian Seung's lab at the Massachusetts Institute of Technology was highlighted. Seung and colleagues have been able to develop a machine-learning algorithm that can help trace the path of individual neurons through the brain (Jain et al., 2006). In Seung's program, a machine "watches" as humans go through and map individual neurons. It then examines how the human researchers did this work and develops parameters to follow the same pattern, therefore potentially providing a tool that would dramatically decrease the number of person hours required to some of the work.

To localize proteins and other chemicals efficiently and construct the neurochemical microcircuitry of the brain will require the equivalent of the automated sequencers that drove, with increasing rapidity, the sequencing of the human genome, said Joseph Coyle, professor of psychiatry and neuroscience at Harvard Medical School.

There is no way that a human mind or a collection of human minds could effectively and efficiently sift through the tremendous amount of data. Rather, it is going to require automated procedures running on computers that have proved themselves in one domain being applied to this domain, added Read Montague, professor of neuroscience at the Human Neuroimaging Lab, Baylor College of Medicine.

Lichtman stressed that this is big science. No single laboratory can do this. Rather it can only be done through a multilaboratory, national, even an international effort.

All of these advances have researchers like Lichtman and Marder very excited.

"I would say, today, 2008, 2009, we are right at a historical cusp, because we have revolutionary opportunities for circuit analysis in the next decade," said Marder.

"Is this a possibility?" asked Lichtman, who used the word "connectome" to refer to the wiring diagram of the brain. "Can we get connectomes? I would argue that we can. Finally, there are the necessary techniques to do this."

The Importance of Neural Networks

The connectome, of course, is just one step, a way of breaking the brain down into understandable pieces. New research shows that the brain is significantly more than the sum of its parts, and that a network-level view is critical to understanding how it functions.

When information comes in from the outside world—say, when you look at the *Mona Lisa*—the sensory input is transformed in the brain into a series of electrical spikes. It is not that one or two neurons fire; entire regions of the brain (and perhaps the entire brain itself) light up, with a complexity of pathways that tells us a simple circuit map cannot fully account for activity in the brain.

William Bialek, a professor at the Joseph Henry Laboratories of Physics and the Lewis-Sigler Institute of Integrative Genomics, Princeton University, described this series of spikes at the workshop as "the language in which the nervous system does its business."

"Although much of the history of neuroscience is about understanding the responses of individual neurons," said Bialek, "in fact, almost all of our experiences are based on the activity of many, many neurons."

He put forward the human retina as an example. If you measure the correlations among different neurons processing information from the retina, you find that the correlations are very weak. Therefore, it is tempting to assume that it is the individual neurons that matter, and not the whole. But Bialek says some order is hiding in the code.

Although all the correlations among neurons are weak, nearly all pairs are correlated. Intriguingly, this is reminiscent of models for how collective opinions form in societies, but it is also reminiscent of earlier models in statistical physics, where, in fact, surprisingly dramatic collective effects can be hiding in these weak correlations.

John Hopfield proposed just such a model of neural networks in 1982, and the model has been supported by the research in many ways. Bialek explained, for instance, that these networks have a tendency to fall into different "states," or general patterns of electrical spikes, which are more consistent than the individual firing of single neurons. If you play a movie to the retina twice, for instance, the exact neurons that fire will change each time. The overall pattern of brain activity, however, will be retained and reproduced.

We have already made great strides in being able to understand these codes, according to some at the workshop. Theodore Berger, professor of Engineering at the University of Southern California noted that multisite recording array technologies and new advances in computer algorithms, including nonlinear dynamic models, have made it much easier to understand the representations of the outside world in the brain. There was the strong suggestion, by Berger, that technological developments would rapidly translate into substantial breakthroughs or developments.

In the past decade or two, we have achieved a great deal in brain mapping and localization per se, but today the need is to move from brain localization to connectivity imaging, remarked He.

Others thought that even more surprising patterns may emerge—patterns we cannot even imagine today.

Montague argued that the field of neuroscience brings psychological concepts of behavior to the table, working with assumptions that the brain works in a particular way and that these assumptions influence how we study the brain.

"When we look for neural correlates—we go look for the neural correlates of learning and memory or we go look for the neural correlates of scratch-pad memory or long-term memory—maybe there are some hidden concepts there that a more agnostic approach on the outside and the inside would reveal," said Montague.

Montague called for more rigorous definitions of behavior and a more agnostic approach to research, using the power of modern computing technology to search for patterns we cannot even imagine. The time is ripe for a bottom-up analysis in which one can move away from psychological space to computational space, with good quantification of behavioral endpoints.

The Way Forward

A true theory of the brain, in some ways, is the ultimate goal: understanding how the physical processes in our neurons turn into behaviors and perceptions of the outside world.

As the above discussions demonstrated, and as summarized by the session chair and Provost of Harvard University, Steven Hyman, we are still in the early stages of answering that question, or even figuring out what that question might look like. There was widespread support in the room for the importance of mapping the physical circuitry of the brain, but there was also a feeling that a physical map alone would not be sufficient to explain how it actually works. There were suggestions to focus on neural networks and the language of electrical activity in the brain, as well as efforts to drive agnostic data crunching to search for patterns that we cannot even imagine.

Panelists generally agreed that great technological breakthroughs have made this effort more possible now than ever before, but that additional breakthroughs—particularly in imaging and computer learning—were needed.

In the end, the payoff from this kind of research would be huge. Not only is developing a viable theory of the brain's capabilities one of the great intellectual challenges in mankind's history, but this research would also have tremendous applications for curing disease, guiding education policies, and maintaining health.

We have reached a technical point where it becomes feasible to imagine approaching an understanding of the way the brain is constructed at a level of detail, granularity, and rigor so that we could imagine that taking shape and reaching a theory of the mind and the brain at some point, commented Dennis Choi, former president of the Society of Neuroscience and the Director of the Comprehensive Neuroscience Center at Emory University. All that remains is to do it.

GRAND CHALLENGE: NATURE VERSUS NURTURE: HOW DOES THE INTERPLAY OF BIOLOGY AND EXPERIENCE SHAPE OUR BRAINS AND MAKE US WHO WE ARE?

Nature vs. nurture is one of the oldest questions in science. The answer is not an either/or, but rather it is both nature and nurture, acting in various degrees.

As summarized below in greater detail, many workshop participants—including Hyman, Marder, and Michael Greenberg, chair of the Department of Neurobiology at Harvard Medical School—chose to highlight the nature versus nurture question as one of the Grand Challenges of the field, but in so doing, they put a twist on the question, asking: *How does the interplay of biology and experience shape our brains and make us who we are?*

The key word there is "interplay." "Interplay" suggests, and modern research in neuroscience demands, that there is a back and forth pattern between nature and nurture, a dynamic system that involves a continuous feedback loop shaping the physical structure of our brains.

Brain Plasticity

Thirty years ago, the working assumption in neuroscience was this: People are born with a set number of neurons, hardwired in a certain way, and brain function is essentially all downhill from there. We spend our lifetimes losing connections and neurons—the brain slowly falling apart until we die.

Except it is not true. In 1998, Fred "Rusty" Gage, working out of the Laboratory of Genetics at the Salk Institute, showed that the human brain can and does produce new nerve cells into adulthood (Eriksson et al., 1998). In mice, he showed that exercise could increase the rate of neurogenesis, showing that the system is not fixed, but responds itself to experience and the outside world. The discovery of neurogenesis and an improved understanding of neuroplasticity—the ability of the brain to shape, form, eliminate, and strengthen new connections throughout life—has completely recast the question of nature versus nurture.

"Neurons can change their connectivity," explained Blakemore. "They can change the strength of their connections. They can change the morphology of their connections. They can do it not necessarily just in early stages of life, although that is especially exaggerated, but probably throughout life responding to new environments and experiences."

New research shows, for instance, that the number and strength of connections we have in the brain is determined by how often those connections are stimulated. The brain, if you will, has a "use it or lose it" approach to neurological maintenance.

Genetic programming also plays a key role. In most cases, the initial formation of a synapse occurs independent of stimulation. But if that synapse is not used, the brain will "prune" or eliminate it. Conversely, the more often a connection is used, the stronger it becomes in a physical sense, with more dendritic spines connecting to one another and a stronger net connection over time.

On the developmental side, researchers now understand the critical role that sensory input plays in shaping the wiring of the brain from the earliest days. Blakemore discussed work in his lab on the development of neural wiring in mice. Researchers have known since the 1960s that the neurons connected to the ultrasensitive whiskers of mice align themselves in a format called "barrel fields." Each of these barrel fields is connected to a single whisker, although how or why they influence function is unknown. Blakemore showed that if you removed a clump of whiskers at an early age, the segment of the brain linked to that area never develops the barrel structure.

Similar research has shown in mice that if you tape one eye shut from birth, the mouse never gains the ability to see from that eye—it needs the stimulation to develop. However, if you tape shut the eye of an adult mouse for a similar period of time, vision is not affected.

All this seems to point the finger toward experience, but of course, the system really works as a complete feedback loop.

"We used to think . . . that the capacity of the brain to change its connections was an entirely independent process from the genetic regulation of structure," said Blakemore. "But, of course, that cannot be the case. If adaptive change is possible, that must be the consequence of having molecular mechanisms that mediate those changes. Plasticity is a characteristic that has been selected for, so there must be genes for plasticity."

In the case of barrel fields, Blakemore's lab and other investigators have identified a number of molecules and genes that appear to be involved in mediating between incoming information for the whiskers and the anatomical changes necessary to produce the barrel field.

Understanding how this interplay works has huge implications for understanding how our brain develops and changes over time, and raises a number of interesting questions. Marder, for instance, asked how the brain can be so plastic and yet still retain memories over time.

Plasticity, however, is just one half of the equation; the underlying genetics are critically important, and new techniques and technologies make this a particularly interesting time to address these questions. For instance, modern, high-throughput gene-profiling technologies allow researchers to figure out all of the underlying transcriptions in a neuron, and see how these are manifest in the body.

Understanding the interplay of biology and experience on learning and development will surely require understanding the biological processes that cause changes in individual neurons and synapses. But this is only part of the puzzle. We must also understand the control of learning processes at a system-wide level in the brain. How does the brain orchestrate the right set of neural synaptic updates based on training experiences we encounter over our lifetime? Given the tremendous number of synapses in the brain, it is unlikely that a purely bottom-up approach will suffice to answer this question.

A complementary approach to studying experience-based learning at a system level relies on machine learning algorithms that have been developed to allow robots to learn from experience, described Mitchell. One intriguing study has shown that temporal-difference learning algorithms, which enable robots successfully to learn control strategies such as how to fly helicopters autonomously, can be used to predict the neural activity of dopamine-based systems in the human brain that are involved in reward-based learning (Schultz et al., 1997; Seymour et al., 2004; Doya, 2008). The integration of such system-level computational models alongside new research into synaptic plasticity offers an opportunity to examine the interplay of biology and experience on learning and development from multiple perspectives.

New tools will allow researchers to understand how variability between different genes and neurons and neuronal activity could influence behavior and capabilities across different people, the researchers said. Who we are is not only influenced by the yes/no expression of genes, but also the specific levels of expression among different genes, which in turn influences neuronal activity.

Gene-Environment Interactions

Nature and nurture are not simply additive interactions that result in a particular behavior, but rather a complex interplay of many factors. Nature includes not only the usual factors—parents, homes, what people learn—but also many other factors that individuals are exposed to routinely in their daily environments. As Marder emphasized, we cannot simply assume that gene X produces behavior Y. Instead as Bialek described, there are often many additional factors that directly and indirectly interact with gene X and ultimately influence variants in behavior. These variants define individuality.

As previously described, it has been known for almost 50 years that experience from the outside environment shapes our brain. This comes initially from the original work of Nobel Laureates David Hubel and Torsten Wiesel who studied how information is sensed and processed in the part of the brain responsible for vision. As Greenberg commented, the field is now at a point where we could in the next 10 years attain a significant mechanistic understanding of how the environment impinges directly on our genes to give rise to a malleable organ that allows us to adapt and change.

Huge Clinical Importance

Multiple participants at the workshop—including Nora Volkow, director of the National Institute on Drug Abuse; Joseph Takahashi, investigator of the Howard Hughes Medical Institute and Northwestern University; Lichtman; and Coyle—highlighted the role of genetics in shaping the brain as one of the fundamental challenges for neuroscience, both for its basic scientific interest and for its practical applications: Understanding how genes and experience come together to impact the brain could significantly alter how we think about treating neurological disease. Many of the most common neurological and mental health disorders—schizophrenia, bipolar disorder, autism, Parkinson's disease, multiple sclerosis, Alzheimer's disease—are complex genetic disorders that are influenced by environmental factors.

Alcino Silva, professor in the Departments of Neurobiology, Psychiatry and Psychology at the University of California, Los Angeles, showcased research from his lab showing he could treat and reverse developmental disorders in adult mice. This finding is worth repeating because it is so contrary to our general thinking on developmental disorders: Scientists working out of Silva's lab have been able to reverse the impacts of the developmental disorder NF-1 (Neurofibromatosis type 1), which is caused by genetic malfunction, by treating the pathology of the disease in adult mice. These mice, which have obvious cognitive deficits, regain mental function when treated; Silva has advanced the study into human clinical trials.

The applications of this vein of study extend beyond developmental disorders. A growing body of evidence is revealing a massive feedback loop among genetics, neurological structure, experience, and disease. You are three times more likely to die from a heart attack if you are depressed than if you are not, for instance, and depression has a huge impact on diabetes as well, stated Coyle.

Taking a step backward, clinical data also show that people who experience multiple stressful episodes in their lives tend to suffer from clinical depression. But there is tremendous variation: Some people are resistant to stress and others are not.

"It turns out that the pattern is correlated with a polymorphic variation in one particular gene, the gene for the transporter for serotonin, a transmitter which is known to be involved in regulating mood," explained Blakemore.

How do genes work in the brain to determine our resilience to stress, and how can those capabilities be monitored and modulated for better health?

The Way Forward

Asking these kinds of questions was not realistic 10 or even 5 years ago. The advent of high-throughput gene profiling and the growing sophistication of our ability to manipulate genes in animal models lets us, for the first time, explore the role that genes play in both creating and modulating our neural structures. At the same time, new imaging techniques and technologies like channel rhodopsin "light switches" let us better characterize neural systems and their response to the world around us, and to begin to plumb the tremendous feedback loop among genes, experience, and the physical activity in the brain.

Until quite recently, these have remained philosophical questions, commented Marder. However, the field of neuroscience is now in a position—through all the molecular, connectomics, and technological advances—to put these questions on firm mechanistic, biological bases, and to attack them scientifically.

GRAND CHALLENGE: HOW DO WE KEEP OUR BRAINS HEALTHY? HOW DO WE PROTECT, RESTORE, OR ENHANCE THE FUNCTIONING OF OUR BRAINS AS WE AGE?

If the percentage of the population facing neurological disease is large, the percentage facing the impacts of aging is total, that is, the aging body and brain impact everyone as they get older. There is no question that the brain changes naturally as it ages—just ask any 50-year-old how often they forget where their car keys are—but there is little understanding of how and why aging causes the brain to change. Understanding the physical changes that occur as the brain ages would be an important place to start in efforts to slow down, eliminate, or reverse the unwanted parts of this process in the future, suggested Volkow.

Questions such as "How does the brain work?" and "How does the interplay of biology and experience shape our brains and make us who we are?" are phenomenally interesting, and have many practical corollaries. But workshop participants, including Timothy Coetzee, executive director of Fast Forward of the National Multiple Sclerosis Society, also recognized that their research aims to have an immediate impact on easing the suffering of those facing neurological disease. It comes down to understanding questions such as: How do we keep our nervous system healthy as we age? Are there ways to protect, restore or enhance the function of our brains with aging?

The Question of Aging

A great deal of neuroscience research is played out against the backdrop of the time bomb of disease, said Blakemore. According to the World Health Organization (WHO), neurological and mental health disorders have a tremendous impact on individuals throughout their lifespan. It is estimated that 10 to 20 percent of children suffer from mental or behavioral problems and one in every four people develops a neurological disorder at some stage in life (WHO, 2001). Therefore, the time lost and economic impact caused by mental and neurological illness is tremendous.

Not only is the scale of the problem enormous, but it is growing as the population ages, and neither the public nor the scientific community is content to wait on basic discovery before we start investigating cures. "Society [has a] hunger for interventions long before we have a deep, fundamental knowledge" of how the neurological system works, said Hyman.

For many aging people, the question of how and why their brains age is much deeper than small forgetfulness; it goes down right to the core of personality. "People are obviously interested in how 'me' is developed," said Marcelle Morrison-Bogorad, director of the Division of Neuroscience at the National Institute on Aging. "But they are also very interested in how 'me' is retained and how to retain 'me' in the presence of aging-related changes which take 'me' away."

Theories do exist. Many have noted a rise in inflammatory markers in the aging brain, and guessed about ischemic effects that build up over time. There are signs of decreases of protein transcription and protein expression in the brain, signal-transduction alterations that likely lead to the morphological changes that are observed, including decreased numbers of neurons and connections between neurons Some believe that a lifetime of toxic exposures play a role, although we have not yet conducted the kind of epidemiological studies that would provide this information.

Starting at Square One

For many neurological disorders, we are really at square one in understanding how a particular disease works, and what avenues we should explore for treatment, let alone having a better understanding of what life style adjustments could be made to avoid or minimize the onset of aging-related complications. Many participants, including Greenberg and Steven Dekosky, chair of the Department of Neurology at the University of Pittsburgh, expressed a desire for a better core understanding of the physical morphology of neurological disease, as well as the physical morphology of aging. The ability to diagnose presymptomatic disease by either looking for biomarkers or, better yet, studying the genetic makeup, genetic expression, and neurological makeup of individual patients would be one good proxy for gaining an understanding of how diseases arise. This kind of research was not possible a few years ago, before the advent of high-throughput genetic sequencing and high-resolution neuro-imaging, but it is becoming increasingly possible every day.

The Complicated Role of Genetics

A popular presumption is that many diseases are driven by a single genetic mutation, and that a magic switch in the body is either on or off—and as a result, you either have a disorder such as autism or Parkinson's or you do not.

Increasingly, however, research suggests that these are disorders of complex genetics, where multiple genes and varying levels of expression are combined to create the impact of the disease. Understanding the etiology of a disorder is further compounded by the influence of the environment on genetic expression. Greenberg explained research showing how parts of the genome are involved in the process of synapse development, synaptic pruning, and the balance between exciting and inhibiting individual synapses.

Another emerging idea is that it is not just a genetic mutation that knocks out function, but subtle mutations that affect the level of expression of the genes and greatly impact disease and normal function. Perhaps this may give us some insight into the processes that lead to "graded" neurological spectrum disorders, such as autism spectrum disorders. As Takahashi highlighted, the use of applications made through advances in genetic tools will allow for a much more integrated understanding of our behaviors. Consequently, an improved understanding of the role of genetics and the environment will almost certainly improve our understanding of how best to protect, restore, or enhance the function of our brains and nervous systems.

The Trouble with Current Treatments

Without a core understanding of how the brain works, the current generation of neurological treatments and preventions is imprecise. In diseases like depression, our best current therapies are to expose the entire brain with a neuromodulator like serotonin, producing only a partial therapeutic response along with unwanted side effects. As Montague observed during the workshop, "We wiggle the knobs down here at the molecular end in a way . . . and we get some sort of behavioral endpoint out there. . . . In between there is nothing."

With depression there is not even a real scientific definition of the focus of the disease—"mood"—and no accurate way to measure how it changes, nor is there a core understanding of how serotonin impacts the brain as a whole to alter mood, explained Coyle.

Similarly, Montague described how there is very limited understanding of the widespread "placebo effect," both in neurological diseases and in other physical diseases. What is the physical morphology of the placebo effect, and how does the body use that to treat and cure itself? The impact is not small; huge efforts are undertaken to account for it in clinical trials. While it appears that the neurotransmitter dopamine has a role in the process, we still do not know how the total process works.

Marder also highlighted deep brain stimulation (DBS), in which electrodes are implanted into the brain to treat Parkinson's disease, depression, and other maladies. Although the process of DBS is based on some understanding of what areas of the brain are impacted by disease, there is no depth to our understanding of the physical process by which DBS works. DBS is a perfect example of where a fundamental understanding of the structure and function of the brain could drive tremendous benefits. New neuroimaging techniques and new neuronal mapping techniques make it easy to imagine mapping out the structure and functional map of the brain in such a way that we could precisely target an intervention like DBS to create a desired treatment effect, for example as has been done to treat individuals suffering from Parkinson's disease.

However, as Volkow described, at the end of the day the brain not only gives rise to some disorders of the nervous system, but it is also where emergent behaviors originate. Consequently, the brain is very likely to be driving the likelihood of optimizing health, through determining our behaviors, which then affects our lifestyles and our health.

INSPIRING THE NEXT GENERATION OF SCIENTISTS

As the previous pages explain, much of the discussion at the workshop focused on the ways that advances in neuroscience can help us treat disease, handle aging, and otherwise improve the health and functioning of the brain. This is the core charge of neuroscientists, and drives many of the Grand Challenges identified during the workshop.

However, who among us is not fascinated by the brain? Who does not want to know how it works, why it fails, how we learn, and how our current personalities develop? These are some of the most fundamentally interesting questions in the world.

The fact that these questions can now be approached in a rigorous way, Leshner hoped, would capture the attention of the public, particularly budding scientists of all ages. Ultimately, answering these questions will take more than just focus and money (although both of those will be important); it will take smarts and effort, two resources that can only be tapped by capturing the imagination of our youth.

Thomas Insel, director of the National Institute of Mental Health, identified this inspiration as a potential Grand Challenge for the field. "I think it would be a fantastic Grand Challenge to have neuroscience taught at every high school in America, that we make this as appealing as astrology might be to the American public—or football," said Insel. "It is something that could . . . [ensure] that this field will move even faster and further than it has in the last 10 years, if that is possible."

CHALLENGES AND TECHNICAL LIMITATIONS

Many barriers that have impeded researchers from addressing the questions highlighted in the Grand Challenges workshop have disappeared over recent years, remarked Leshner. Advances in imaging technology, new techniques such as those similar to the Brainbow, and neuronal "light switches" have laid the groundwork for researchers to explore the brain as never before. However, many of the advances that have been made over the last decade have also been a direct result of basic unrestricted discovery research. For example, the increased use and power of the internet and computer programming, sequencing the human genome, and the discovery of small non-coding RNA, are all examples of the value of basic discovery research that have had major impact on how we view and understand our brains and nervous systems. It is very likely that future unexpected discoveries and advances in other areas of physics, biochemistry, computer science, and molecular biology will continue to have a significant impact on the future progress that will be made in the neurosciences.

But the path from where we are today to where we want to go is not easy. Both conceptual and technical impediments must be solved. This document does not intend to capture each and every one of those challenges—the science is too intricate and involved—but rather to highlight a few high-level topics raised by multiple workshop participants.

Integrating Neuroscience and Working Toward a Common Goal

"Grand Challenges" are designed to unite a scientific field around a few common problems. This is not easy. The nature of science is that researchers are often focused on micro-fine topics and must promote the importance of their particular corner of expertise to secure funding and attention for their fields. The result can be scientific fiefdoms and intellectual turf wars, emboldened by the need for financial support.

The problem is more acute in neuroscience than in other fields. As mentioned earlier, one of neuroscience's great strengths is also its greatest weakness: It is not a single "science" at all, but an interdisciplinary field drawing on biology, chemistry, computer science, genetics, and others.

"It is a very large continuum . . . from molecular to behavioral neuroscience, with extraordinary opportunities," said Story Landis, director of the National Institute of Neurological Disorders and Stroke at the National Institutes of Health (NIH). "We need to figure out how to portray the excitement across that continuum in a way that not only the public and our funders, but, most important, the neuroscience community as a whole, can embrace."

Working with Psychological Concepts and Defining Behavior

A further challenge highlighted by some at the workshop was to free neuroscience from its roots in psychology and psychiatry. "We have either enjoyed or suffered under the concepts that psychology has brought to us for the last, let's say, 100 years," said Montague.

A further challenge highlighted by some at the workshop was the need to reconcile understanding of psychological phenomenon at the behavioral and cognitive levels with understanding at the molecular and cellular levels. Montague decried the disconnection between much of cognitive neuroscience and molecular neuroscience. Terms like perception, awareness, consciousness and disease states like depression, anxiety, mood, do not easily translate into their underlying molecular mechanism. Therefore, we need to find a common language that allows both ends of the neuroscience spectrum to communicate. This likely will require an agreement on a common unit of analysis, which is the most reduced unit for the cognitive and most complex for the molecular neuroscience approaches.

Montague, Hyman, and others argued for the need for more concrete and quantitative definitions of behavior-understanding behavior derived from an agnostic approach to the problem, rather than one driven by our preconceived ideas about how the brain functions."

New Technological Requirements

Despite tremendous advances in the past few years, many workshop participants highlighted the need for additional technical advances to drive the field forward. Although the workshop did not focus too closely on specific technological needs, one technology stood out: an imaging device or series of devices that can offer both ultra-fine spatial imaging resolution and ultra-fine time resolution.

Techniques are needed that can produce both high resolution in space and high resolution in time, said Blakemore. Magnetic resonance imaging (MRI) and positron emission tomography (PET) provide fairly good resolution in space, but they are slow techniques. Electrical recording from the brain with an EEG or MEG can give us high temporal resolution, but poor spatial resolution. Finding ways of combining these characteristics of different techniques, or new kinds of methodology, which can provide improved spatial and temporal resolution in both is going to be very important for the future.

Professor He agreed, emphasizing that the tool must be noninvasive to study the human brain. He added that the field needed a new way to image connectivity in the brain and that such a tool would have major clinical appeal as well. "[Y]ou can help with the surgical planning on epilepsy patients," said He. "You can help treat a lot of neurological disease by rationally designing a neuromodulation or neurostimulation paradigm if you know the pattern. You just block that pattern and you can treat the patient even without surgery."

ETHICAL CONSIDERATIONS

The brain is an object of great fascination and power. It is the seat of humanity, the source of everything we are and everything we want to be.

Understanding how the brain works—really understanding, on a core physiological level—would have tremendous benefits for society. But it would also raise significant moral, ethical, and practical considerations, which neuroscience must address carefully as it moves forward.

"I think it is useful to realize that neuroscientists do operate in a kind of interestingly sensitive area," said Moreno. "As the old saying goes, just because you are not paranoid doesn't mean somebody is not following you."

As Moreno explained, people become nervous when they hear questions such as "How does the brain work?" and how to intervene in the brain. "The idea that scientists can have what I call technologically mediated access [to the brain]—can use devices or drugs, fancy machines that most of us do not really understand . . . I think is of great concern to many people and is something that, going forward, the community needs to think about," said Moreno.

A comparison was made to the Human Genome Project, which attracted a great deal of concern from both the public and professional ethicists because it edged so closely to the foundations of life. Ultimately, extensive education and careful restrictions convinced people that the genome project was a safe idea, but only because its backers addressed the topic directly and in a public manner. Understanding "how the brain works" raises similar issues, and must be discussed, examined, and considered in the same light.

Clinical Concerns

Moreno raised a number of additional areas where ethics should impact the work of researchers. For instance, in clinical trials, it is possible that neurological interventions could change people's sense of themselves. How can these kinds of changes be measured, monitored, and understood, not only as they happen, but in the process of obtaining informed consent and in the investigative process itself? In a similar vein, Moreno pointed out, as we develop a better understanding of the presymptomatic risk factors for certain diseases, the issue of how to notify research subjects of their likelihood of developing neurological disease becomes a major concern. This is already a live challenge in diseases such as the debilitating and deadly Huntington's disease, which can be diagnosed in a presymptomatic state, but is invariably fatal. Expanding capabilities to identify disease on a presymptomatic basis would expand the potential treatment options of these challenges exponentially.

Fostering a Dialog

Throughout the discussion, ethical and morals concerns were raised. The ongoing discussion of learning disabilities, for instance, and the potential to intervene and mediate disorders medically, caused concern among many on the question of streamlining and mainstreaming in education and the cost to society of losing diversity within the population. Similarly, discussions of transhumanism—supercharging the brain—made some hesitant, while others saw it as a means to help the elderly regain function.

The overarching point was that neuroscience stands on the cusp of huge advances, and those huge achievements raise major issues that the field has never considered before.

"The time is really now to start thinking about what that means and how we want to . . . self-regulate and engage in better professional forethought as to how the impact of what we are doing inside our laboratories is actually reaching beyond the borders of our community," said Insel.

The community was acutely aware that if they do not self-regulate their efforts and engage the public in a focused dialogue on the issue of neuroscience, politicians and other nonscientists will do it for them.

CONCLUSION

The purpose of a forum at the National Academies is not to come to consensus or make specific recommendations to the public. It is, rather, to foster an open discussion among leading experts in the field; to gather some of the best and brightest around a common topic and see what emerges.

To that end, Leshner proclaimed the workshop a tremendous success. The opportunity to step back and discuss the big issues surrounding neuroscience pulled researchers out of their particular areas of focus and forced them to take a 30,000-foot view of the space. They made it clear that the neurosciences have advanced tremendously over the past 50 years. The progress of the past in combination with new tools and techniques has positioned neuroscience on the cusp of even greater transformational progress in our understanding of the brain and how its activities result in mental activity.

On the Cusp

Neuroscience is on the cusp of exciting breakthroughs that take advantage of the convergence of scientific knowledge and technologies, like Brainbows, neuronal light switches, and computer learning technologies have made it possible to answer questions such as the following:

- How does the brain work and produce mental activity? How does physical activity in the brain give rise to thought, emotion, and behavior?
- How does the interplay of biology and experience shape our brains and make us who we are?
- How do we keep our brains healthy? How do we protect, restore, or enhance the functioning of our brains as we age?

As highlighted during the last panel discussion with Coetzee, Marder, Hyman, Insel, Leshner, Volkow, and Ting Kai Li, director of the National Institute on Alcohol Abuse and Alcoholism, if there was any debate about the feasibility of answering these questions, there was no debate on this: Doing so would have tremendous benefits to society, easing the suffering of those with disease, helping people age gracefully, and even improving our understanding of issues like learning disabilities and more. It is a classic investment problem—taking money away from the current need to invest in a brighter future, commented Coetzee. However, the advantages gained from understanding the mechanisms of brain function, plasticity, and other topics would lead to step-wise improvements in therapies—improvements that cannot happen any other way.

The challenges will be great, said Landis. Integrating the various fields of neuroscience toward a common goal will be tough, and the field still requires new technological advances and ideas to achieve its goals. It will be a step-wise process, with the benefits taking years or even decades to realize. But with the right injection of new funding and resources, there was a feeling that the potential payoff balanced this load.

"Both NIH and NSF believe that the field [of neuroscience] is now poised on a threshold of major transformational advances," said Olsen. "I do not think it is an exaggeration to say that . . . in the next decade and beyond 'neuro' will become the new 'nano' in terms of experimental capabilities that are beyond anything we could previously imagine, and discoveries that fire the imagination, achieve great practical advances, and grow the economy."

Added Olsen: "I think the potential benefits are too enormous to let this opportunity pass."

I would say, today, 2008, 2009, we are right at a historical cusp. . . .

—Eve Marder

A

References

Boyden, E. S., F. Zhang, E. Bamberg, G. Nagel, and K. Deisseroth. 2005. Millisecond-timescale, genetically-targeted optical control of neural activity. *Nature Neuroscience* 8(9):1263–1268.

Doya, K. 2008. Modulators of decision making. *Nature Neuroscience* 11(4):410–416.

Eriksson, P. S., E. Perfilieva, T. Björk-Eriksson, A. M. Alborn, C. Nordborg, D. A. Peterson, and F. H. Gage. 1998. Neurogenesis in the adult human hippocampus. *Nature Medicine* 4(11):1313–1317.

Jain, V., V. Zhigulin, and H. S. Seung. 2006. Representing part-whole relationships in recurrent neural network. *Adv Neural Info Proc Syst* 18:563–570.

Kelly, R. C., M. A. Smith, J. A. Samonds, A. Kohn, A. B. Bonds, J. A. Movshon, and T. S. Lee. 2007. Comparison of recordings from microelectrode arrays and single electrodes in the visual cortex. *The Journal of Neuroscience* 27(2):261–264.

Livet, J., T. A. Weissman, H. Kang, R. W. Draft, J. Lu, R. A. Bennis, J. R. Sanes, and J. W. Lichtman. 2007. Transgenic strategies for combinatorial expression of fluorescent proteins in the nervous system. *Nature* 450(7166):56–62.

NAE (National Academy of Engineering). 2008. *Grand challenges for engineering*, http://www.engineeringchallenges.org/ (accessed August 28, 2008).

NRC (National Research Council). 2003. *Connecting quarks with the cosmos: Eleven science questions for the new century*. Washington, DC: The National Academies Press.

NSF (National Science Foundation). 2006. *Grand challenges of the mind and brain*. Arlington, VA.

Schultz, W., P. Dayan, P. R. Montague. 1997. A neural substrate of prediction and reward. *Science* 275(5306):1593–1599.

Seymour, B., J. P. O'Doherty, P. Dayan, M. Koltzenburg, A. K. Jones, R. J. Dolan, K. J. Friston, and R. S. Frackowiak. 2004. Temporal difference models describe higher-order learning in humans. *Nature* 429(6992):664–667.

WHO (World Health Organization). 2001. *The world health report 2001—Mental health: New understanding, new hope.* Geneva, Switzerland.

B

Workshop Agenda

From Molecules to Minds:
Challenges for the 21st Century

June 25, 2008

Lecture Room
The National Academy of Sciences Building
2101 Constitution Avenue, N.W.
Washington, DC 20418

Background:

The unifying theme for this workshop is the need to expand the understanding of how perception, cognition, and action arise in the human brain from interactions among molecules, chemicals, neurons, and circuits, the brain's fundamental building blocks. This concept is pertinent to every level of brain organization, from understanding how molecules become assembled into neurons to how neurons get assembled into neuronal circuits, how those circuits develop unique properties and capabilities, and finally how dysfunction at any of these levels may lead to disorders of the brain.

Objectives:

- Illuminate the progress and successes made by the neuroscience community and highlight the challenges still facing the field.

- Identify the guiding principles, fundamental scientific questions, and goals that will inspire the scientific and public communities to support and engage this grand challenge.
- Identify the infrastructure and resource requirements that will be necessary to advance and accelerate discovery, including:

  o   What will be the technology needs?
  o   Which disciplines will need to be engaged and what will be their training requirements?
  o   What partnerships need to be forged?

9:00 a.m.   Welcome, Introductions, and Workshop Objectives

> ALAN LESHNER, *Forum Chair*
> Executive Publisher
> *Science Magazine*
> Chief Executive Officer
> American Association for the Advancement of Science

9:15 a.m.   Overview and Objectives of the IOM Neuroscience Forum's Grand Challenges Initiative

> KATHIE OLSEN
> Deputy Director
> National Science Foundation

9:30 a.m.   From Molecules to Minds: Opportunities and Challenges

> COLIN BLAKEMORE
> Chief Executive Officer (former)
> British Medical Research Council
> Professor
> Department of Physiology, Anatomy, and Genetics
> Oxford University

SESSION I:
OVERVIEW OF CURRENT KNOWLEDGE:
EXAMINING THE CURRENT THEORIES OF HOW
THE NERVOUS SYSTEM IS ORGANIZED FROM
MOLECULES TO MINDS

Session Objective: Highlight and discuss the current understanding, hypotheses, and theories for how the nervous system is organized, and how molecular and cellular organization impacts the function of the brain. Based on the current understanding, what are the future needs for the neuroscience community?

9:50 a.m.   Introduction to the Session: Session Objectives

> STORY LANDIS, *Session Chair*
> Director
> National Institute of Neurological Disorders and Stroke

10:00 a.m.   Principles of Neuronal Coding

> WILLIAM BIALEK
> John Archibald Wheeler/Battelle Professor in Physics
> Joseph Henry Laboratories of Physics
> Lewis–Sigler Institute for Integrative Genomics
> Princeton University

10:15 a.m.   Circuits: Between Systems and Cellular Processes

> EVE MARDER
> Professor of Neuroscience
> Department of Biology and Volen Center
> Brandeis University

10:30 a.m.   BREAK

10:45 a.m.   Cognitive Disorders: Molecules, Cells, and Circuits

ALCINO SILVA
Director
Behavioral Testing Core
Professor, Psychology
Tennenbaum Center for the Biology of Creativity,
  Neurobiology
UCLA School of Medicine

11:00 a.m.   Research and Neuroethics

JONATHAN MORENO
David and Lyn Silfen University Professor
Center for Bioethics
University of Pennsylvania Health System

11:15 a.m.   Discussion

STORY LANDIS, *Session Chair*
Director
National Institute of Neurological Disorders and Stroke

Noon   LUNCH

SESSION II:
EXPLORING THE GRAND CHALLENGES

Session Objective: Highlight and discuss cross-cutting themes and knowledge gaps, and how these may help to identify a set of guiding principles and fundamental scientific questions.

- What the field knows it does not know;
- What it doesn't know it doesn't know; and
- What it thinks it knows but doesn't.

1:00 p.m.   Introduction to the Session: Session Objectives

STEVEN HYMAN, *Session Chair*
Provost
Harvard University

1:15 p.m.   Panel Discussion 1: What are the current challenges and opportunities?

Machine Learning and Its Implications on Understanding the Brain

TOM MITCHELL
Fredkin Professor of AI and Machine Learning
Chair
Machine Learning Department
School of Computer Science
Carnegie Mellon University

Neurophysiologic and Modeling Strategies to Understand the Brain

THEODORE BERGER
David Packard Professor of Engineering, Professor of Biomedical Engineering and Neurobiology
Director
Center for Neural Engineering
University of Southern California

Computational Neuroscience: What Lies Ahead?

READ MONTAGUE
Professor
Department of Neuroscience
Human Neuroimaging Lab
Baylor College of Medicine

2:00 p.m.   Panel Discussion 2: What are the current challenges and opportunities?

Understanding Neuronal Connections Using Imaging Strategies

JEFF LICHTMAN
Professor
Molecular and Cellular Biology
Harvard University

Multimodal Neuroimaging of Brain Activity and Connectivity

BIN HE
Professor of Biomedical Engineering, Electrical Engineering, and Neuroscience
Interim Director, Center for Neuroengineering
University of Minnesota

Molecular Neurobiology and Genetics of Circadian Clocks

JOSEPH S. TAKAHASHI
Investigator
Howard Hughes Medical Institute
Walter and Mary E. Glass Professor in the Life Sciences
Department of Neurobiology and Physiology
Northwestern University

Pruning the Brain Through Changes in Activity

MICHAEL GREENBERG
Chair
Department of Neurobiology
Harvard Medical School

3:00 p.m.   BREAK

3:15 p.m. Panel Discussion 3: What are the current challenges and opportunities?

Neurochemistry and the Brain

> JOSEPH COYLE
> Eben S. Draper Professor of Psychiatry and of Neuroscience
> Harvard Medical School

The Aging Mind: Structural and Neurochemical Changes

> STEVEN T. DEKOSKY
> Professor and Chair
> Department of Neurology
> University of Pittsburgh

SESSION III:
NEXT STEPS: ENERGIZING THE COMMUNITY

Session Objective: What "grand challenges" were identified during the workshop that will inspire the scientific and public communities to support and engage this initiative? Identify and discuss current and future technological and resource needs that will be necessary to overcome associated challenges and advance and accelerate discovery. How can we, and who should, champion the innovation and ideas discussed during the workshop?

4:15 p.m. Panel Discussion with Key Stakeholders: Opportunities, Priorities, and Resources Requirements Identified During the Workshop

> TIMOTHY COETZEE
> Vice President, Discovery Partnerships
> National Multiple Sclerosis Society

EVE MARDER
Professor of Neuroscience
Department of Biology and Volen Center
Brandeis University

STEVEN HYMAN
Provost
Harvard University

TOM INSEL
Director
National Institute of Mental Health

ALAN LESHNER
Executive Publisher
Science Magazine
Chief Executive Officer
American Association for the Advancement of Science

TING-KAI LI
Director
National Institute on Alcohol Abuse and Alcoholism

NORA VOLKOW
Director
National Institute on Drug Abuse

5:00 p.m.　Closing Remarks

KATHIE OLSEN
Deputy Director
National Science Foundation

5:15 p.m.　ADJOURN

C

Registered Workshop Attendees

Cara Allen
National Institute of
  Neurological Disorders and
  Stroke (NINDS), National
  Institutes of Health (NIH)

Peter Basser
Principal Investigator
Eunice Kennedy Shriver
National Institute of Child
  Health and Human
  Development, NIH

Jennifer Belding
Intern
American Psychological
  Association

Arden Bement
Director
National Science Foundation

William Cheetham
General Electric (GE)

Semahat Demir
Program Director
Biomedical Engineering
Program
National Science Foundation

Diane DiEuliis
Senior Policy Analyst
Office of Science and
  Technology Policy

Juan Dominguez
Professor
University of Texas

Irene Farkas-Conn

Brian Ferrar
First Secretary S&I
British Embassy

Gregory Gratson
Program Manager
GE Global Research

Giovanna Guerrero
Health Policy Analyst
NINDS, NIH

Marlene Guzman
Senior Advisor to the Director
National Institute of Mental
 Health (NIMH), NIH

Peggy Hanson
Neurobehavioral Consultants

Erin Heath
Senior Policy Associate
American Association for the
 Advancement of Science

Elizabeth Hoffman
Legislative and Federal Affairs
 Officer
American Psychological
 Association

Michael Huerta
Associate Director
NIMH, NIH

Amanda Johnson
American Psychological
 Association

Susan Koester
Deputy Director
Division of Neuroscience and
 Basic Behavioral Science,
 NIMH, NIH

Raaj Mehta
Center for the Study of the
 Presidency

Richard Nakamura
Deputy Director and Acting
 Scientific Director
NIMH, NIH

Molly Oliveri
Director
Division of Developmental
 Translational Research,
 NIMH, NIH

Sangeeta Panicker
Director of Research Ethics
American Psychological
 Association

Lisa Pfeifer
Graduate Student
Guest Researcher
University of Maryland and
 NIH

Oxana Pickeral

Lorenzo Refolo
Program Director
NINDS, NIH

Paul Scott
Director
Office of Science Policy and
 Planning, NINDS, NIH

Neelam Shah
Kaiser Permanente

Philip Wang
Director
Division of Services and
  Intervention Research, NIMH,
  NIH

John Yeh

Robert Zalutsky
Senior Science Advisor
NINDS, NIH

D

Biographical Sketches of Invited Speakers, Planning Committee Members, Forum Members, and Staff

INVITED SPEAKERS

Alan I. Leshner, Ph.D. (*Forum Chair*), is chief executive officer of the American Association for the Advancement of Science (AAAS) and executive publisher of its journal, *Science*. Previously Dr. Leshner had been director of the National Institute on Drug Abuse (NIDA) at the National Institutes of Health (NIH), and deputy director and acting director of the National Institute of Mental Health (NIMH). Before that, he held a variety of senior positions at the National Science Foundation (NSF). Dr. Leshner began his career at Bucknell University, where he was a professor of psychology. Dr. Leshner is an elected member (and on the governing council) of the Institute of Medicine (IOM), and a Fellow of AAAS, the National Academy of Public Administration, and the American Academy of Arts and Sciences. He was appointed by the U.S. President to the National Science Board, and is a member of the Advisory Committee to the Director of NIH. He received an A.B. in Psychology from Franklin and Marshall College and an M.S. and a Ph.D. in physiological psychology from Rutgers University. Dr. Leshner has been awarded six honorary Doctor of Science degrees.

Theodore Berger, Ph.D., is the David Packard Professor of Engineering and Professor of Biomedical Engineering and Neuroscience, and director, Center for Neural Engineering (CNE) at the University of Southern California. Dr. Berger received his Ph.D. in Physiological Psychology from Harvard University in 1976, and continued postdoctoral training in the Psychobiology Department at the University of California, Irvine, and at the Salk Institute for Biological Studies. His current research focuses on developing biologically based mathematical models of the functional properties of the hippocampus by combining experimental studies of fundamental electrophysiological properties of hippocampal neurons and theoretical studies based on nonlinear systems and compartmental analyses, and experimental studies of cellular/molecular mechanisms of synaptic plasticity and the effects of such plasticity on functional dynamics of the hippocampus at the network and systems level. Through collaborations with other CNE faculty, Dr. Berger's research extends to developing analog VLSI implementations of experimentally based models of hippocampal neurons and neural networks, both for basic research and applications, and developing "neuron–silicon interface" technology using silicon-based, multisite electrode arrays and tissue culture methods for implantation of hardware models into the brain to replace damaged or dysfunctional nerve tissue.

William Bialek, Ph.D., is the John Archibald Wheeler/Battelle Professor in Physics at Princeton University. He is also an associated faculty member in the Department of Molecular Biology, and a member of the multidisciplinary Lewis–Sigler Institute. Professor Bialek participates in the interdepartmental educational programs in Applied and Computational Mathematics, Biophysics, Neuroscience, and Quantitative and Computational Biology. Dr. Bialek attended the University of California, Berkeley, receiving the A.B. (1979) and Ph.D. (1983) degrees in Biophysics. After postdoctoral appointments at the Rijksuniversiteit Groningen in the Netherlands and at the Institute for Theoretical Physics in Santa Barbara, he returned to Berkeley to join the faculty in 1986. He joined the Princeton faculty as a professor of physics in 2001. Professor Bialek's research interests have covered a wide variety of theoretical problems at the interface of physics and biology, from the dynamics of individual biological molecules to learning and cognition. Best known for contributions to our understanding of coding and computation in the brain, Dr. Bialek and collaborators have shown that aspects of brain function can be described as essentially optimal strategies for adapting to the complex dynamics of the world, making the most of the available signals in the face of fundamental physical constraints and limitations.

Colin Blakemore, Ph.D., studied Medical Sciences in Cambridge University from 1962 to 1965 and completed a Ph.D. in Physiological Optics as a Harkness Fellow at the University of California in 1968. Since 2003 he has been on leave while holding the post of chief executive of the British Medical Research Council. He has maintained research activity at

Oxford and since October 2007 he has held the title of professor of neuroscience. Dr. Blakemore is a fellow of the Royal Society and the Academy of Medical Sciences, is an Honorary FRCP and holds honorary fellowships from the Institute of Biology and the British Association for the Advancement of Science. He is a foreign member of the Royal Netherlands Academy of Arts and Sciences. He has been president of the British Association for the Advancement of Science, the British Neuroscience Association, the Physiological Society, and the Biosciences Federation. Dr. Blakemore's research has been concerned with many aspects of vision, early development of the brain, and plasticity of the cerebral cortex. His current interests are in two areas. First, together with Dr. Irina Bystron, he is studying the earliest stages of formation of the cerebral cortex in human embryos, using immunocytochemical methods and techniques for tracing the outgrowth of axons to examine the proliferation of neural stem cells; the production, migration, and differentiation of cortical neurons; as well the formation of connections into and out of the developing cortex. One aim of this research is to define the developmental errors that underlie cognitive disorders, such as autism, dyslexia, and schizophrenia. His second area of current research, together with Drs. Kai Thilo and Meng Liang, uses techniques for imaging activity in the living adult human brain to examine the capacity of sensory areas of the cortex to reorganize their activity during selective attention, during the integration of information from different sensory systems, and after the onset of blindness.

Timothy Coetzee, Ph.D., is the executive director of Fast Forward, LLC, a venture philanthropy of the National Multiple Sclerosis Society. He is responsible for the Society's strategic funding of biotechnology and pharmaceutical companies as well as partnerships with the financial and business communities. Prior to assuming his current position, Dr. Coetzee led the Society's translational research initiatives on nervous system repair and protection in multiple sclerosis (MS) as well as the Society's programs to recruit and train physicians and scientists in MS research. Dr. Coetzee received his Ph.D. in Molecular Biology from Albany Medical College in 1993 and has since been involved in MS research. He was a research fellow in the laboratory of Society grantee Dr. Brian Popko at the University of North Carolina–Chapel Hill, where he received an Advanced Postdoctoral Fellowship Award from the Society. After completing his training with Dr. Popko, Dr. Coetzee joined the faculty of the Department of Neuroscience at the University of Connecticut School of Medicine, where he conducted research that applied new technologies to understand how myelin is formed in the nervous system. He is the author of a number of research publications on the structure and function of myelin. Dr. Coetzee joined the National MS Society's Home Office staff in fall 2000.

Joseph Coyle, M.D., holds the Eben S. Draper Chair of Psychiatry and Neuroscience at Harvard Medical School. From 1991 to 2001, he served as chairman of the Consolidated Department of Psychiatry at Harvard Medical School, which included the nine hospital programs of psychiatry affiliated with the Medical School. After graduating from Holy Cross College, he received his M.D. from Johns Hopkins School of Medicine in 1969. Following an internship in Pediatrics, he spent 3 years at NIH as a research fellow in the laboratory of Nobel laureate Julius Axelrod, Ph.D. He returned to Hopkins in 1973 to complete his residency in psychiatrics, in which he is board certified, and joined the faculty in 1975. In 1980, he was promoted to professor of Neuroscience, Pharmacology, and Psychiatry; in 1982 he assumed the directorship of the Division of Child and Adolescent Psychiatry, being named the Distinguished Service Professor in 1985. Dr. Coyle's research interests include developmental neurobiology, mechanisms of neuronal vulnerability, and psychopharmacology. In particular, he has carried out research on the role of glutamatergic neurons in the pathophysiology of neuropsychiatric disorders for 30 years. He has a long commitment to training. In the mid-1980s, he was the principal investigator (PI) of a NIMH Training Grant that had a core curriculum, which introduced Ph.D. fellows to psychiatric illnesses with patient demonstrations. While president of the Society of Neuroscience, he worked with NIMH to develop a minority training grant. For the past 10 years, he has served as co-PI on this grant, which was the foundation for minority mentoring and networking for the Society. He has published more than 500 scientific articles and has edited seven books. His research has been cited more than 35,000 times, and his H-factor is 93. He has received continuous NIH funding for his research for 30 years and is the director of an NIMH Conte Center on the Neurobiology of Schizophrenia (2001–2011). Dr. Coyle is a member of the IOM (1990), a Fellow of the American Academy of Arts and Sciences (1993), a Distinguished Fellow of the American Psychiatric Association, a Fellow of the American College of Psychiatry, and a Fellow of AAAS. He served on the National Advisory Mental Health Council for NIMH from 1990 to 1994. He is past president of the American College of Neuropsychopharmacology (2001) and past president (1991) of the Society for Neuroscience, which has more than 35,000 members. He sits on more than 20 journal editorial boards, including *JAMA*, and is editor-in-chief of the *Archives of General Psychiatry*, the most highly cited journal in the field (citation impact: 13.9).

Steven DeKosky, M.D., is professor and chair of the Department of Neurology and director of the Alzheimer's Disease Research Center at the University of Pittsburgh. His clinical research includes differential diagnosis, neuroimaging, and genetic risks for Alzheimer's disease and trials of new medications. His basic research centers on structural and neurochemical changes in human brains in normal aging and dementia. He is director of a national multicenter trial to assess whether Ginkgo biloba can delay onset of dementia in normal elderly adults. In 2004 he was appointed to the Peripheral and Central Nervous System Advisory Committee of the Food and Drug Administration (FDA). Dr. DeKosky was a member of the national board of directors of the Alzheimer's Association for 8 years, the last 4 as vice chair. He was the chair of the Alzheimer's Association Medical and Scientific Advisory Council from 1997 to 2002. He also chairs the Professional Advisory Board of the Greater Pittsburgh Chapter and was a founding member of the Lexington–Blue Grass Chapter of the Alzheimer's Association. In 2002 he was elected chair of the Medical and Scientific Advisory Panel of Alzheimer's Disease International, the international organization of national Alzheimer's associations. Dr. DeKosky has served as chair of the Section on Geriatrics of the American Academy of Neurology (AAN). He chaired the recent AAN Practice Parameters Committee for Early Detection, Diagnosis, and Management of Dementia. He has been an examiner in neurology for the American Board of Psychiatry and Neurology (ABPN) for more than 15 years and served as a member of the ABPN Part I (Written) Examination Committee for 10 years. In 2002 he was elected to the Neurology Council of the ABPN; he is one of the eight neurologists who oversee board certification in the country. He has received a Teacher Investigator Development Award from the National Institute of Neurological Disorders and Stroke (NINDS) and the Presidential Award of the American Neurological Association, and is listed in "The Best Doctors in America." He has published more than 200 peer-reviewed articles and book chapters. Dr. DeKosky's academic career began in 1978 at the University of Virginia, School of Medicine in Charlottesville, where he was an instructor in the Department of Neurology until 1979 and completed a postdoctoral fellowship in neurochemistry at the Clinical Neuroscience Research Center in the Department of Neurology. Before joining the UPMC in 1990, Dr. DeKosky was on the faculty of the University of Kentucky, College of Medicine for 11 years, serving as codirector of the Alzheimer's Disease Research Center and interim chair of Neurology for 2 years. From 1992 to 2000, he was head of the Division of Geriatrics and Neuropsychiatry in the Department of Psychiatry at the University of Pittsburgh and Western Psychiatric Institute and Clinic, where he holds a joint appointment as professor of psychiatry. After earning a bachelor's degree in Psychology from Bucknell University, PA. Dr. DeKosky attended the University of Florida–Gainesville for graduate studies in psychology and neuroscience. In 1974, Dr. DeKosky graduated from the University of Florida College of Medicine. After an internship in internal medicine at the Johns Hopkins Hospital, Dr. DeKosky completed a 3-year residency in neurology at the University of Florida in 1978.

Michael Greenberg, Ph.D., is the director of the Division of Neurosciences at Children's Hospital Boston and professor of neurology and neurobiology at Harvard Medical School. Dr. Greenberg received his Ph.D. from The Rockefeller University and completed a postdoctoral fellowship in molecular biology at that institution. He subsequently completed a fellowship in molecular biology at New York University Medical Center. He holds the F. M. Kirby Foundation Neuroscience Directorship at Children's Hospital Boston. Research in the Greenberg lab has focused on identifying the mechanisms by which extracellular stimuli trigger cellular responses that are critical for proliferation, differentiation, and survival of cells in the developing nervous system and for the adaptive responses of neurons in the mature nervous system.

Bin He, Ph.D., is a professor in the Department of Biomedical Engineering at the University of Minnesota. Dr. He received his B.S. in Electrical Engineering from Zhejiang University in 1982, and his Ph.D. in Biomedical Engineering with the highest honors from Tokyo Institute of Technology, a Nobel Prize–winning campus in 1988. He completed the postdoctoral fellowship in biomedical engineering at Harvard University–Massachusetts Institute of Technology (MIT). After working as a research scientist at MIT, he was on faculty of the Departments of Electrical Engineering and Bioengineering at the University of Illinois–Chicago, where he was named a University Scholar by the president of the University of Illinois. Since January 2004, he has been a professor of biomedical engineering, electrical engineering, and neuroscience, and director of the Biomedical Functional Imaging and Neuroengineering Laboratory at the University of Minnesota. Dr. He's research interests include functional biomedical imaging, neuroengineering, cardiovascular engineering, and bioelectromagnetism.

Steven Hyman, M.D., is provost of Harvard University and a professor of neurobiology at Harvard Medical School. From 1996 to 2001, he served as NIMH Director. Earlier, he was a professor of psychiatry at Harvard Medical School; director of psychiatry research at Massachusetts General Hospital; and the first faculty director of Harvard University's Mind, Brain, and Behavior Initiative. In the laboratory he studied the regulation of gene expression by neurotransmitters, especially dopamine, and by drugs that influence dopamine systems. This research was aimed at understanding addiction and the action of therapeutic psychotropic drugs. Dr. Hyman is a member of the IOM, a fellow of the American Academy of Arts and Sciences, and a fellow of the American College of Neuropsychopharmacology. He is editor-in-chief of the *Annual Review of Neuroscience* and has won public service awards from the federal government and patient advocacy groups such as the National Alliance for the Mentally Ill and the National Mental Health Association. Dr. Hyman received his B.A., *summa cum laude*, from Yale College in 1974, and an M.A. from the University of Cambridge, which he attended as a Mellon Fellow studying the history and philosophy of science. He earned his M.D., *cum laude*, from Harvard Medical School in 1980.

Thomas R. Insel, M.D., graduated from Boston University, where he received a B.A. from the College of Liberal Arts and an M.D. from the Medical School. He did his internship at Berkshire Medical Center, Pittsfield, MA, and his residency at the Langley Porter Neuropsychiatric Institute at the University of California, San Francisco. In 1979 Dr. Insel joined NIMH, where he served in various scientific research positions until 1994, when he became as a professor, Department of Psychiatry, Emory University School of Medicine, and director of the Yerkes Regional Primate Research Center. As director of Yerkes, Dr. Insel built one of the nation's leading HIV vaccine research programs. He also served as the founding director of the Center for Behavioral Neuroscience, a Science and Technology Center, funded by NSF to develop an interdisciplinary consortium for research and education at eight Atlanta colleges and universities. Dr. Insel's scientific interests have ranged from clinical studies of obsessive-compulsive disorder to explorations of the molecular basis of social behaviors in rodents and nonhuman primates. His research on oxytocin and affiliative behaviors, such as parental care and pair bonding, helped to launch the field of social neuroscience. Dr. Insel oversees the NIMH's $1.4 billion research budget, which supports to investigators at universities throughout the country in the areas of basic science; clinical research, including large-scale trials of new treatments; and studies of the organization and delivery of mental health services.

Story C. Landis, Ph.D., has been director of NINDS since September 2003. Dr. Landis oversees an annual budget of $1.5 billion and a staff of more than 900 scientists, physician–scientists, and administrators. The Institute supports research by investigators in public and private institutions across the country, as well as by scientists working in its intramural laboratories and branches in Bethesda, MD. Since 1950, the Institute has been at the forefront of U.S. efforts in brain research. Dr. Landis joined the NINDS in 1995 as scientific director and worked with then-Institute Director Zach W. Hall, Ph.D., to coordinate and reengineer the Institute's intramural research programs. Between 1999 and 2000, under the leadership of NINDS Director Gerald D. Fischbach, M.D., she led the movement, together with NIMH Scientific Director Robert Desimone, Ph.D., to bring a sense of unity and common purpose to 200 laboratories from 11 NIH Institutes, all of which conduct leading-edge clinical and basic neuroscience research. A native of New England, Dr. Landis received her undergraduate degree in Biology from Wellesley College in 1967 and her master's degree (1970) and Ph.D. (1973) from Harvard University, where she conducted research on cerebellar development in mice. After postdoctoral work at Harvard University studying transmitter plasticity in sympathetic neurons, she served on the faculty of the Harvard Medical School Department of Neurobiology. In 1985 she joined the faculty of Case Western Reserve University School of Medicine in Cleveland, OH. She held many academic positions there, including associate professor of pharmacology; professor and director of the Center on Neurosciences; and chair of the Department of Neurosciences, a department she was instrumental in establishing. Under her leadership, Case Western's Neurosciences Department achieved worldwide acclaim and a reputation for excellence. Throughout her research career, Dr. Landis has made many fundamental contributions to the understanding of developmental interactions required for synapse formation. She has garnered many honors and awards and is an elected Fellow of the Academy of Arts and Sciences, AAAS, and the American Neurological Association. In 2002, she was named president-elect of the Society for Neuroscience.

Ting-Kai (TK) Li, M.D., earned his undergraduate degree from Northwestern University and his M.D. from Harvard University, and completed his residency training at Peter Bent Brigham Hospital in Boston, where he was named chief medical resident in 1965. He also conducted research at the Nobel Medical Research and Karolinska Institutes in Stockholm and served as deputy director of the Department of Biochemistry within the Walter Reed Army Institute of Research. Dr. Li joined the faculty at Indiana University as professor of medicine and biochemistry in 1971. He was subsequently named the school's John B. Hickam Professor of Medicine and Professor of Biochemistry and later Distinguished Professor of Medicine. In 1985 he became director of the Indiana Alcohol Research Center at the Indiana University School of Medicine, where he also was the associate dean for research. Dr. Li is the recipient of numerous awards for his scientific accomplishments, including the Jellinek Award, the James B. Isaacson Award for Research in Chemical Dependency Diseases, and the R. Brinkley Smithers Distinguished Science Award. Dr. Li has also served in many prominent leadership and advisory positions, including past president of the Research Society on Alcoholism, and as a member of the National Advisory Council on Alcohol Abuse and Alcoholism and the Advisory Committee to the Director, NIH. Dr. Li was elected to membership in the IOM in 1999 and is also an honorary Fellow of the United Kingdom's Society for the Study of Addiction.

Jeff Lichtman, Ph.D., is professor of molecular and cellular biology at Harvard University. Dr. Lichtman's interests lie in the mechanisms that underlie synaptic competition among neurons that innervate the same target cell. Such competitive interactions are responsible for sharpening the patterns of neural connections during development and may also be important in learning and memory formation. His laboratory studies synaptic competition by visualizing synaptic rearrangements directly in living animals using modern optical imaging techniques. They have concentrated on neuromuscular junctions in a very accessible neck muscle in mice where new transgenic animals and other labeling strategies allow individual nerve terminals and postsynaptic specializations to be monitored over hours or months. In addition, they have developed several new methods to improve the ability to resolve synaptic structure.

Eve Marder, Ph.D., is the Victor and Gwendolyn Beinfield Professor of Neuroscience in the Biology Department and Volen Center for Complex Systems at Brandeis University. She received her Ph.D. in 1974 from UCSD (the University of California, San Diego), and subsequently conducted a one-year postdoctoral at the University of Oregon, then a 3-year postdoc at the Ecole Normale Superieure in Paris. She became an assistant professor in the Biology Department at Brandeis University in 1978, and was promoted to professor in 1990. During her time at Brandeis, Dr. Marder has been instrumental in the establishment of undergraduate and graduate programs in neuroscience. She is a fellow of AAAS, a fellow of the American Academy of Arts and Sciences, and a trustee of the Grass Foundation. She was the Forbes Lecturer at the MBL in 2000 and the Einer Hille Lecturer at the University of Washington in 2002. Dr. Marder has studied the dynamics of small neuronal networks using the crustacean stomatogastric nervous system. Her work was instrumental in demonstrating that neuronal circuits are not "hardwired," but can be reconfigured by neuromodulatory neurons and substances to produce a variety of outputs. Together with Larry Abbott, her laboratory pioneered the "dynamic clamp." Marder was one of the first experimentalists to forge long-standing collaborations with theorists, and has for nearly 15 years combined experimental work with insights from modeling and theoretical studies. Her work today focuses on understanding how stability in networks arises despite ongoing channel and receptor turnover and modulation, both in developing and adult animals.

Tom Mitchell, Ph.D., is the E. Fredkin Professor and head of the Machine Learning Department at Carnegie Mellon University. His research interests are generally in machine learning, artificial intelligence, and cognitive neuroscience. Dr. Mitchell is author of the widely used textbook *Machine Learning*. He is past chair of the AAAS Section on Information, Computing, and Communication. He is past president of the American Association of Artificial Intelligence (AAAI), and a recent member of the National Research Council's Computer Science and Telecommunications Board. Dr. Mitchell's recent research includes the use of machine learning and brain imaging (fMRI) to study the neural representation of word meanings in the human brain. One recent result showed that the neural activity representing meanings of concrete nouns such as "hammer" and "apartment" are quite similar across different people, as demonstrated by the fact that a classifier trained on a group of people would successfully decode the item being considered by new people. A second recent result proposed a computer model that predicts the fMRI neural activity representation for arbitrary concrete nouns, based on statistics of the noun's use in a trillion-word collection of online text.

Read Montague, Ph.D., is a professor in the Department of Neuroscience at Baylor College of Medicine, where he is also director of the Human Neuroimaging Lab and director of the Center for Theoretical Neuroscience. His work focuses on computational neuroscience—the connection between the physical mechanisms present in real neural tissue and the computational functions that these mechanisms embody. Work in the Montague group also extends into several experimental areas, including synaptic physiology, human neuroimaging, and human behavior. The Montague lab is also a member of The Computational Psychiatry Unit, a new unit dedicated to understanding the computational connections between biological mechanisms and psychiatric illness

Jonathan Moreno, Ph.D., is the David and Lyn Silfen University Professor and professor of medical ethics and of history and sociology of science at the University of Pennsylvania. He comes to Penn in connection with the Penn Integrates Knowledge (PIK) initiative, a University-wide initiative launched in 2005 by Penn President Amy Gutmann to recruit exceptional faculty members whose research and teaching exemplify the integration of knowledge across disciplines. Dr. Moreno holds a joint appointment in HSS (School of Arts and Sciences) and in medical ethics in the School of Medicine. He is also a senior fellow at the Center for American Progress in Washington, DC and a visiting professor of biomedical ethics at the University of Virginia. From 1998 to 2006, Dr. Moreno held the Emily Davie and Joseph S. Kornfeld Chair in Biomedical Ethics at the University of Virginia. Dr. Moreno is an elected member of the IOM and has been a member of numerous National Academies committees. He co-chaired the Committee on Guidelines for Human Embryonic Stem Cell Research. He has served as a senior staff member for two Presidential advisory committees and has given invited testimony for both houses of Congress.

Kathie L. Olsen, Ph.D., became deputy director of the National Science Foundation in August 2005. She joined NSF from the Office of Science and Technology Policy in the Executive Office of the President, where she was the associate director and deputy director for science, responsible for overseeing science and education policy, including physical, life, environmental, behavioral, and social sciences. Earlier, she served as chief scientist at the National Aeronautics and Space Administration (NASA) from May 1999 to April 2002 and as acting associate administrator for the new Enterprise in Biological and Physical Research (July 2000–March 2002). As NASA chief scientist, she served not only as the NASA Administrator's senior scientific advisor and principal interface with the national and international scientific communities, but was also the principal adviser to the Administrator on budget content of the scientific programs. Before joining NASA, Dr. Olsen was the senior staff associate for the Science and Technology Centers in the NSF Office of Integrative Activities. Other work experience includes serving as a legislative fellow at the Brookings Institute, working on an NSF detail in a Senator's office, and serving as acting deputy director for the NSF Division of Integrative Biology and Neuroscience. Dr. Olsen received her B.S. with honors from Chatham College in Pittsburgh, where she majored in Biology and Psychology and was elected to Phi Beta Kappa. She earned her Ph.D. in Neuroscience at the University of California–Irvine. She was a postdoctoral fellow in the Department of Neuroscience at Children's Hospital of Harvard Medical School. Subsequently, at the State University of New York (SUNY)–Stony Brook, she was both a research scientist at Long Island Research Institute and assistant professor in the Department of Psychiatry and Behavioral Science at the Medical School. Her research on neural and genetic mechanisms underlying development and expression of behavior was supported by the NIH. Her awards include the NSF Director's Superior Accomplishment Award; the International Behavioral Neuroscience Society Award; the Society for Behavioral Neuroendocrinology Award for outstanding contributions in research and education; the Barry M. Goldwater Educator Award from the American Institute of Aeronautics and Astronautics–National Capital Section; and NASA's Outstanding Leadership Medal. She has also received honorary degrees from Chatham College, Clarkson University, and University of South Carolina.

Alcino Silva, Ph.D., is a professor in the Departments of Neurobiology, Psychiatry, and Psychology head of the Center for the Biology of Creativity; and coordinator of Learning and Memory at the Brain Research Institute at the University of California, Los Angeles. The laboratory uses a number of approaches to study learning and memory, enhanced cognitive function, and creativity. Although many of the studies include mice with defined mutations, the laboratory has incorporated a number of approaches that these problems to be addressed at multiple levels of biological complexity. The lab uses slice electrophysiology to study the cell biology of learning and memory, single-unit recordings to tap into the properties of circuits during learning, and neuroanatomic lesions to assess the role of specific brain regions. It also uses time-lapse, two-photon confocal imaging in vivo to define cellular neurosystems and dendritic structural changes during remote memory.

Joseph Takahashi, Ph.D., is an investigator at Howard Hughes Medical Institute and the Walter and Mary E. Glass Professor in the Life Sciences in the Department of Neurobiology and Physiology at Northwestern University. Dr. Takahashi leads an effort to sort out the genes and proteins that make up the mammalian circadian clock—and then to figure out how they work together ultimately to control the behavior of the animal. In 1997 Dr. Takahashi's lab discovered the first circadian rhythm gene in mammals, the mouse gene *Clock*, an acronym for "circadian locomotor output cycles kaput." Since then, researchers have identified eight other mammalian circadian genes, enabling scientists to study the interplay of proteins that make up the clock's negative-feedback mechanism at the most basic molecular level.

Nora D. Volkow, M.D., became director of the National Institute on Drug Abuse in May 2003. Dr. Volkow came to NIDA from Brookhaven National Laboratory (BNL), where she held concurrent positions, including associate director for life sciences, director of nuclear medicine, and director of the NIDA–Department of Energy Regional Neuroimaging Center. In addition, Dr. Volkow was a professor in the Department of Psychiatry and associate dean of the Medical School at SUNY–Stony Brook. Dr. Volkow brings to NIDA a long record of accomplishments in drug addiction research. She is a recognized expert on the brain's dopamine system, with her research focusing on the brains of addicted, obese, and aging individuals. Her studies have documented changes in the dopamine system affecting the actions of frontal brain regions involved with motivation, drive, and pleasure and the decline of brain dopamine function with age. Her work includes more than 350 peer-reviewed publications, 3 edited books, and more than 50 book chapters and non-peer-reviewed manuscripts. The recipient of multiple awards, she was elected to membership in the IOM and was named "Innovator of the Year" in 2000 by *U.S. News and World Report*. Dr. Volkow received her B.A. from Modern American School, Mexico City; her M.D. from the National University of Mexico, Mexico City; and her postdoctoral training in psychiatry at New York University. In addition to BNL and SUNY–Stony Brook, Dr. Volkow has worked at the University of Texas Medical School and Sainte Anne Psychiatric Hospital in Paris.

PLANNING COMMITTEE MEMBERS

Alan I. Leshner, Ph.D., (*Forum Chair*) see Speaker bio.

Alan Breier, M.D., is a professor of psychiatry at the Indiana University School of Medicine. He was named vice president of medical and chief medical officer for Eli Lilly and Company in August 2003. He is a member of the Lilly Research Laboratories (LRL) Policy Committee and Lilly's Senior Management Council. He joined Lilly as an LRL Research Fellow in March 1997, the same year he was appointed adjunct professor of psychiatry at Indiana University School of Medicine in Indianapolis. He received his M.D. from the University of Cincinnati School of Medicine, and trained in psychiatry at Yale University School of Medicine. Dr. Breier was associate research professor of psychiatry at the University of Maryland School of Medicine and chief of the Section on Clinical Studies at the NIMH Intramural Research Program. At Eli Lilly, he has focused on neuroscience drug development and led the Zyprexa Product Team. As chief medical officer, Dr. Breier leads Lilly's medical organization, which annually conducts Phase I through IV clinical trials in more than 60 countries. Dr. Breier has sponsored the Principles of Medical Research, which encompasses ethical standards for medical research, and established Lilly's clinical trial registry, which is a publicly accessible website for posting the initiation and results of clinical trials. Dr. Breier has received several awards, including the A.E. Bennett Neuropsychiatric Research Foundation Award and the Joel Elkes International Award. He is a fellow of the American College of Neuropsychopharmacology and has published more than 225 scientific papers. He is included in Best Doctors in America.

David H. Cohen, Ph.D., is a professor of psychiatry and biological sciences at Columbia University, where served as vice president and dean of the Faculty of Arts and Sciences from 1995 to 2003. Prior to joining Columbia, he served as vice president for research and dean of the graduate school and subsequently as provost at Northwestern University. He has held professorships in physiology and/or neuroscience at Northwestern, SUNY–Stony Brook, University of Virginia School of Medicine, and Case Western University School of Medicine. Dr. Cohen has held various elected offices in national and international organizations, including president of the Society for Neuroscience and chair of the Association of American Medical Colleges. He has served on varied boards including, for example, Argonne National Laboratory, Fermi National Accelerator Laboratory, Zenith Electronics, and Columbia University Press. He has also served on numerous advisory committees for various organizations, including NIH, NSF, Department of Defense, and The National Academies Dr. Cohen received his B.A. from Harvard University and Ph.D. from the University of California, Berkeley, and was an NSF postdoctoral fellow at the University of California, Los Angeles.

Richard Hodes, M.D., is the director of the National Institute on Aging (NIA) at the NIH. He maintains an active involvement in NIH research through his direction of the Immune Regulation Section, a laboratory studying regulation of the immune system, focused on cellular and molecular events that activate the immune response. In the past Dr. Hodes acted as a clinical investigator for the National Cancer Institute (NCI), then as deputy chief and acting chief of the NCI Immunology Branch. Since 1982 he has served as program coordinator for the U.S.–Japan Cooperative Cancer Research Program, and since 1992 he has been on the scientific advisory board of the Cancer Research Institute. He is also a diplomate of the American Board of Internal Medicine. In 1995 Dr. Hodes was elected as a member of The Dana Alliance for Brain Initiatives; in 1997 he was elected to be an AAAS Fellow; and in 1999 he was elected to membership in the IOM. Dr. Hodes received his M.D. from Harvard Medical School. He completed a research fellowship at the Karolinska Institute in Stockholm and clinical training in internal medicine at the Massachusetts General Hospital.

Steve Hyman, M.D., see Speaker bio.

Judy Illes, Ph.D., is professor of neurology and Canada Research Chair in Neuroethics for the National Core for Neuroethics at the University of British Columbia. Dr. Illes received her doctorate in Hearing and Speech Sciences from Stanford University in 1987, with a specialization in Experimental Neuropsychology. Dr. Illes returned to Stanford University in 1991 to help build the research enterprise in imaging sciences in the Department of Radiology. She also cofounded the Stanford Brain Research Center (now the Neuroscience Institute at Stanford), and served as its first executive director between 1998 and 2001. Most recently Dr. Illes was acting associate professor of pediatrics (medical genetics) and director of the program in neuroethics at the Stanford Center for Biomedical Ethics. Dr. Illes has written numerous books and edited volumes and articles. She is author of *The Strategic Grant Seeker: Conceptualizing Fundable Research in the Brain and Behavioral Sciences* (1999, LEA Publishers, NJ); special guest editor of *Topics of Magnetic Resonance Imaging*, "Emerging Ethical Challenges in MR [Magnetic Resonance] Imaging" (2002); and an article in *Brain and Cognition*, "Ethical Challenges in Advanced Neuroimaging" (2002). Her latest book, *Neuroethics: Defining the Issues in Theory, Practice and Policy*, was published by Oxford University Press in 2006. Dr. Illes is co-chair of the Committee on Women in Neuroscience for the Society for Neuroscience; a member of the Internal Advisory Board for the Institute of Neurosciences, Mental Health and Addiction of the Canadian Institutes of Health Research; and a member of the Dana Alliance for Brain Initiatives.

Thomas R. Insel, M.D., see Speaker bio.

Story C. Landis, Ph.D., see Speaker bio.

Ting-Kai (TK) Li, M.D., see Speaker bio.

Michael D. Oberdorfer, Ph.D., is director of the Strabismus, Amblyopia and Visual Processing, and Low Vision and Blindness Rehabilitation Programs at the National Eye Institute of the NIH. He is involved in a number of trans-NIH initiatives and activities in neuroscience and other areas, including the Coordinating Committee of the NIH Blueprint for Neuroscience Research. Previously, he was a program officer at the NSF, where he was involved in a number of activities, including directing the Developmental Neuroscience Program. Prior to that he was on the faculty of the University of Texas Medical School in Houston. He received his B.A. at Rockford College and his Ph.D. in Zoology and Neuroscience at the University of Wisconsin–Madison.

Kathie L. Olsen, Ph.D., see Speaker bio.

William Z. Potter, M.D., Ph.D., is vice president, Franchise Integrator Neuroscience at Merck Research Laboratories. Prior to joining Merck, he served as the executive director and Lilly Clinical Research Fellow of the Neuroscience Therapeutic Area at Lilly Research Laboratories. He developed a Lilly/Indiana University fellowship early in 1996 and was named professor of psychiatry at Indiana University Medical Center. Before being associated with LRL, he held the position of chief of the Section on Clinical Pharmacology, Intramural Research Program at the NIMH. He had been with the Public Health Service and the NIH since 1971. He has authored more than 200 publications in the field of preclinical and clinical pharmacology, mostly focused on drugs used in affective illnesses and methods for evaluating drug effects in humans. He has received many honors during his career, including the 1975–1977 Falk Fellow, American Psychiatric Association; 1986 Meritorious Service Medal, Public Health Service; and in 1990, St. Elizabeth's Residency Program Alumnus of the Year Award.

Robert C. Richardson, Ph.D., is the Floyd Newman Professor of Physics and Vice Provost for Research at Cornell University. His past experimental work focused on the study of physical phenomena at very low temperatures. He shared the 1996 Nobel Prize in Physics with David Morris Lee and Douglas Osheroff for their discovery of superfluidity in helium-3. He earned a B.S. in 1958 and an M.S. in 1960 from Virginia Tech. He received his Ph.D. from Duke University in 1965. Dr. Richardson was elected to the National Academy of Sciences in 1986.

Paul A. Sieving, M.D., Ph.D., became director of the National Eye Institute at the NIH in 2001. At the University of Michigan Medical School, he was the Paul R. Lichter Professor of Ophthalmic Genetics and was the founding director of the Center for Retinal and Macular Degeneration in the Department of Ophthalmology and Visual Sciences. Dr. Sieving is known internationally for studies of human progressive blinding genetic retinal neurodegenerations, termed retinitis pigmentosa, and rodent models of these conditions. His laboratory study of pharmacological approaches to slowing degeneration in transgenic animal models led to the first human clinical therapy trial of ciliary neurotrophic factor for retinitis pigmentosa, which he reported in *The Proceedings of the National Academy of Sciences in* 2006. He also successfully treated a genetic mouse model of X-linked retinoschisis using gene transfer, which restored retinal function in adult mice. He maintains a clinical practice for patients with these and other genetic forms of retinal diseases, including Stargardt juvenile macular degeneration. Dr. Sieving served as vice chair for clinical research for the Foundation Fighting Blindness from 1996 to 2001. He serves on the Bressler Vision Award Committee and on the jury for the annual 1 million euro Award for Vision Research of the Champalimaud Foundation, Portugal. He received his M.D. in 1978 and a Ph.D. in Bioengineering in 1981 from the University of Illinois, and completed an ophthalmology residency at the University of Illinois Eye and Ear Infirmary. He was elected to membership in the American Ophthalmological Society in 1993 and the Academia Ophthalmologica Internationalis in 2005. He received an honorary Doctor of Science from Valparaiso University in 2003. He was named as one of The Best Doctors in America in 1998, 2001, and 2005. Dr. Sieving has received numerous awards, including the RPB Senior Scientific Investigator Award, 1998; the Alcon Award, Alcon Research Institute, 2000; and the 2005 Pisart Vision Award from the New York Lighthouse International for the Blind. Dr. Sieving was elected to the IOM in 2006.

Rae Silver, Ph.D., is the Helene L. and Mark N. Kaplan Professor of Natural and Physical Sciences and holds joint appointments at Barnard College and at Columbia University. Dr. Silver is a Fellow of the American Academy of Arts and Sciences, American Association of Arts and Sciences. She has participated extensively in scientific and educational activities, including serving as chair for NASA's Research Maximization and Prioritization Committee reviewing Scientific Priorities for the International Space Station; Society for Neuroscience Program committee (Theme E—Autonomic and Limbic System); chair, External Advisory Committee, NSF Center for the Study of Biological Rhythms at the University of Virginia; search committee member for editors of journals, department chairs, and provost at various institutions; and panel member of a number of committees, including NASA: International Space Station Cost and Management Evaluation Task Force. She has also been a member of the NSF Center for Behavioral Neuroscience External Advisory Board for Georgia State, Emory, and other colleges; a member of the Society for Neuroscience Education Committee Ford Foundation

Minority Fellowship Review panel; and president of Society Research in Biological Rhythms. As senior adviser at the National Science Foundation, she worked with NSF staffers in all the scientific directorates to create a series of workshops to examine opportunities to make advances in neuroscience in the next decade through the joint efforts of biologists, chemists, educators, mathematicians, physicists, psychologists, and statisticians. Dr. Silver's studies of the biological clock in the suprachiasmatic nucleus of the brain were the first to conclusively demonstrate that this brain tissue can be readily transplanted and restore function at a very high success rate in an animal model. The laboratory is renowned for analysis of the input, output, and intraneuronal circuits underlying the function of the brain's master clock. A second line of research entails the study of mast cells (renowned for their role in producing allergic reactions) in modulating brain function and as a major source of brain histamine. The research has been supported without interruption by the NIH and NSF, among other sources. Dr. Silver is deeply committed to educating undergraduate and graduate students both at the national and institutional levels and in the hands-on context of the laboratory. Consistent with this interest, she created the undergraduate program in quantitative reasoning at Barnard College, and published, with colleagues, studies of mathematical learning. She initiated the undergraduate major in neuroscience, serving as its first program director. She also served as director of the graduate program in psychology at Columbia University.

Roy E. Twyman, M.D., is vice president, franchise development in the Central Nervous System/Pain Area of Johnson & Johnson Pharmaceutical Research and Development. He oversees licensing and acquisition efforts for neurology, psychiatry, and pain franchises while coordinating strategic activities for central nervous system (CNS) discovery optimization, early human studies and proof of concept, new technologies, and cross-company projects. Additional oversight includes the pharmacogenomics and neuroimaging teams that support broad-based pharmaceutical research and development (R&D) across all therapeutic areas. Previously, Dr. Twyman was on the faculty at the University of Utah and the University of Michigan. He received his B.S. in Electrical Engineering from Purdue University. He earned his M.D. from the University of Kentucky and completed a neurology residency at the University of Michigan.

Nora D. Volkow, M.D., became Director of the National Institute on Drug Abuse (NIDA) in May, 2003. Dr. Volkow came to NIDA from Brookhaven National Laboratory (BNL), where she held concurrent positions including associate director for life sciences, director of nuclear medicine, and director of the NIDA-Department of Energy Regional Neuroimaging Center. In addition, Dr. Volkow was a professor in the department of psychiatry and associate dean of the medical school at the State University of New York (SUNY)–Stony Brook. Dr. Volkow brings to NIDA a long record of accomplishment in drug addiction research. She is a recognized expert on the brain's dopamine system with her research focusing on the brains of addicted, obese, and aging individuals. Her studies have documented changes in the dopamine system affecting the actions of frontal brain regions involved with motivation, drive, and pleasure and the decline of brain dopamine function with age.

Her work includes more than 350 peer-reviewed publications, three edited books, and more than 50 book chapters and non-peer reviewed manuscripts. The recipient of multiple awards, she was elected to membership in the Institute of Medicine in the National Academy of Sciences and was named "Innovator of the Year" in 2000 by *U.S. News and World Report*. Dr. Volkow received her B.A. from Modern American School, Mexico City, Mexico, her M.D. from the National University of Mexico, Mexico City, and her postdoctoral training in psychiatry at New York University. In addition to BNL and SUNY–Stony Brook, Dr. Volkow has worked at the University of Texas Medical School and Sainte Anne Psychiatric Hospital in Paris, France.

FORUM MEMBERS

Alan I. Leshner, Ph.D., (*Forum Chair*) see Speaker bio.

Huda Akil, Ph.D., is the Gardner Quarton Distinguished University Professor of Neuroscience and Psychiatry at the University of Michigan, and the codirector of the Molecular and Behavioral Neuroscience Institute. Dr. Akil has made seminal contributions to the understanding of the neurobiology of emotions, including pain, anxiety, depression, and substance abuse. Early on, she focused on the role of the endorphins and their receptors in pain and stress responsiveness. Dr. Akil's scientific contributions have been recognized with numerous honors and awards. These include the Pacesetter Award from NIDA in 1993, and with Dr. Stanley

Watson, the Pasarow Award for Neuroscience Research in 1994. In 1998, she received the Sachar Award from Columbia University and the Bristol Myers Squibb Unrestricted Research Funds Award. Dr. Akil is the past president of the American College of Neuropsychopharmacology (1998) and the past president of the Society for Neuroscience (2004). She was elected to be an AAAS fellow in 2000. In 1994, she was elected to membership in the IOM and is currently a member of The National Academies' National Research Council. She was elected to the American Academy of Arts and Sciences in 2004.

Marc Barlow joined the Strategic Marketing group in GE Healthcare as leader of the neuroscience area in 2005. He is responsible for the development and delivery of disease area strategies for CNS. Before joining GE Mr. Barlow was the marketing director of Sanofi-Aventis in the United Kingdom. Previously he held a number of senior sales and marketing positions within the pharmaceutical industry in the United States and abroad. Much of his neuroscience experience focuses on epilepsy, Alzheimer's disease, and stroke. Mr. Barlow graduated from the University of Wolverhampton in 1983 with a focus in Biological Sciences and the Chartered Institute of Marketing with a diploma in Marketing Studies in 1987.

Daniel J. Burch, M.D., is executive vice president of R&D and chief medical officer of CeNeRx Biopharma. He was appointed to his current position in 2007. Dr. Burch has spent his 15-year career in the pharmaceutical industry at Abbott Laboratories, SmithKlineBeecham, and GlaxoSmithKline, where his most recent post was senior vice president, Neurosciences Medicines Development Centre. Dr. Burch holds an M.D. from Vanderbilt University and an M.B.A. from the Wharton School, University of Pennsylvania. He completed a residency in Internal Medicine at Vanderbilt University School of Medicine and a fellowship in Infectious Diseases at Washington University School of Medicine.

Dennis W. Choi, M.D., Ph.D., is executive director of Emory University's Strategic Neurosciences Initiative and director of the Comprehensive Neuroscience Center in the Woodruff Health Sciences Center at Emory. In 1991 he joined Washington University Medical School as head of the Neurology Department; there he also established the Center for the Study of Nervous System Injury and directed the McDonnell Center for Cellular and Molecular Neurobiology. From 2001 until 2006, he was executive vice president for neuroscience at Merck Research Labs. He is an AAAS Fellow and a member of the IOM, the Executive Committee of the Dana Alliance for Brain Research, and the College of Physicians of Philadelphia. He has served as president of the Society for Neuroscience, vice president of the American Neurological Association, and chairman of the U.S./Canada Regional Committee of the International Brain Research Organization. He has also served on the National Academy of Sciences' Board on Life Sciences, and Councils for NINDS, the Society for Neuroscience, the Winter Conference for Brain Research, the International Society for Cerebral Blood Flow and Metabolism, and the Neurotrauma Society. He has been a member of advisory boards for the Christopher Reeve Paralysis Foundation, Grass Foundation, Hereditary Disease Foundation, Spinal Muscular Atrophy Foundation, Harvard–MIT Program in Health Sciences and Technology, Queen's Neuroscience Institute in Honolulu, Max Planck Institute in Heidelberg, Korea Institute for Advanced Study in Seoul, and the FDA, as well as for several university-based research consortia, biotechnology companies, and pharmaceutical companies. He graduated from Harvard College in 1974, and received his M.D. and Ph.D. in 1978 (the latter in Pharmacology) from Harvard University and the Harvard–MIT Program in Health Sciences and Technology. After completing his residency and fellowship training in Neurology at Harvard, he joined the faculty at Stanford University and began research into the mechanisms underlying pathological neuronal death.

Timothy Coetzee, Ph.D., see Speaker bio.

David H. Cohen, Ph.D., see Planning Committee bio.

Richard Frank, M.D., Ph.D., is vice president of clinical and medical strategy at GE Healthcare, Princeton, NJ. He has two decades of experience in designing and implementing clinical trials in the pharmaceutical industry, and built the Experimental Medicine Department at Pharmacia before joining GE Healthcare in 2005. Dr. Frank earned M.D. and Ph.D. (Pharmacology) degrees concurrently and joined the pharmaceutical industry upon completion of his clinical training in 1985. He is past president and founding director of the Society of Non-invasive Imaging in Drug Development and a Fellow of the Faculty of Pharmaceutical Medicine, Royal College of Physicians. He serves on the scientific review board for the Institute for the Study of Aging and is a member of the editorial board of *Molecular Imaging and Biology.*

Richard Hodes, M.D., see Planning Committee bio.

Steven Hyman, M.D., see Speaker bio.

Judy Illes, Ph.D., see Planning Committee bio.

Thomas R. Insel, M.D., see Speaker bio.

Story C. Landis, Ph.D., see Speaker bio.

Ting-Kai (TK) Li, M.D., see Speaker bio.

Husseini K. Manji, M.D., F.R.C.P.C., is vice president, CNS & pain, Johnson & Johnson Pharmaceutical Research and Development. Previously Dr. Manji served as chief of the Laboratory of Molecular Pathophysiology, NIMH, and director of the NIMH Mood and Anxiety Disorders Program, the largest program of its kind in the world. He is also a visiting professor in the Departments of Psychiatry at Columbia University and Duke University. Dr. Manji received his B.S. in Biochemistry and M.D. from the University of British Columbia. Following psychiatry residency training, he completed fellowship training in Psychopharmacology at the NIMH and obtained extensive additional training in Cellular and Molecular Biology at the National Institute of Diabetes and Digestive and Kidney Disorders. The major focus of his ongoing research is the investigation of disease- and treatment-induced changes in gene and protein expression profiles that regulate cellular plasticity and resilience in mood disorders. In broad terms, his laboratories' scientific goals are to capitalize on recent insights into our understanding of the signaling pathways mediating the effects of mood stabilizers in order to understand the pathophysiology of severe mood disorders and to develop improved therapeutics. He has received ongoing research funding for his work on signaling pathways, plasticity, and new medication development for severe mood disorders. Dr. Manji has received numerous research awards, including the A. E. Bennett Award for Neuropsychiatric Research, the Ziskind-Somerfeld Award for Neuropsychiatric Research, the National Alliance for Research in Schizophrenia and Affective Disorders (NARSAD) Mood Disorders Prize (Nola

Maddox Falcone Prize), the Mogens Schou Distinguished Research Award, the American College of Neuropsychopharmacology (ACNP) Joel Elkes award for distinguished research, the Canadian Association of Professors in Psychiatry Award, the Henry and Page Laughlin Distinguished Teacher Award, the Brown University School of Medicine Distinguished Researcher Award, and the NIMH award for excellence in clinical care and research. In addition to his research endeavors, Dr. Manji is also actively involved in medical and neuroscience education endeavors, and has served as a member of the National Board of Medical Examiners Behavioral Science Test Committee, numerous national curriculum committees, the Howard Hughes Medical Institute Research Scholars Program Selection and Advisory Committee, and NIMH's promotion and tenure committee. He developed and directs the NIH Foundation for the Advanced Education in the Sciences graduate course in the Neurobiology of Mental Illness, and has received both the NIMH Mentor of the Year and Supervisor of the Year awards. He has published extensively on the molecular and cellular neurobiology of severe mood disorders and their treatments, has authored numerous textbook chapters, and has edited a book on the mechanisms of action of antibipolar treatments. He is a Fellow of the ACNP, chairs the ACNP's Task Force on New Medication Development, and is a member of the ACNP's Credentialing Committee. He is a member of the NARSAD Scientific Advisory Committee, National Alliance for the Mentally Ill Center on Practice & Research Advisory Committee, the Child and Adolescent Bipolar Foundation Professional Advisory Council, and the Scientific Advisory Board of the Juvenile Bipolar Research Foundation. Dr. Manji is editor of *Neuropsychopharmacology Reviews: The Next Generation of Progress*, deputy editor of *Biological Psychiatry*, associate editor of the journal *Bipolar Disorders*, and a member of the editorial board of numerous journals.

Michael D. Oberdorfer, Ph.D., see Planning Committee bio.

Kathie L. Olsen, Ph.D., see Speaker bio.

Atul Pande, M.D., is senior vice president, Neurosciences Medicines Development Center at GlaxoSmithKline. Dr. Pande received his medical training in India and trained in psychiatry in India and subsequently at the University of Western Ontario in Canada. Following a mood disorders research fellowship at the University of Michigan Medical School,

Dr. Pande served on the Department of Psychiatry faculty. In 1992, Dr. Pande joined the Lilly Research Laboratories in Indianapolis. Since then he has continued his career in pharmaceutical research and has held positions at Parke-Davis Pharmaceutical Research (now part of Pfizer), Pfizer Global R&D, and CeNeRx Biopharma. Dr. Pande has drug development and regulatory submission experience in a broad range of psychiatric and neurological disorders. Dr. Pande has more than 50 peer-reviewed publications, 6 patents, and numerous book chapters, abstracts, and scientific presentations to his credit. Dr. Pande is a member of the Society of Biological Psychiatry, and a Fellow of the Royal College of Physicians and Surgeons of Canada, the American Psychiatric Association, the Canadian College of Neuropsychopharmacology, and the Collegium Internationale Neuro-Psychopharmacologicum.

Menelas Pangalos, Ph.D., is vice president, Neuroscience Research, at Wyeth Research in Princeton, NJ. He previously served as group director and head of Neurodegenerative Research at GlaxoSmithKline in Harlow, United Kingdom. He presents widely on a broad range of topics at international symposia on subjects ranging from strategies for the novel treatment of Alzheimer's disease and GABAB receptor molecular pharmacology to challenges in neuroscience drug discovery. Dr. Pangalos is on the editorial board of *Molecular and Cellular Neuroscience*, on the advisory board for the Wolfson Centre for Age Related Diseases (London University), and has served on the BBSRC Molecular and Cell Biology committee in the United Kingdom. He is also a member of the American Society of Neuroscience and British Pharmacological Society, and an associate of the Royal College of Science. Dr. Pangalos has edited the book *Understanding G-protein Coupled Receptors in the CNS* and has published more than 60 peer-reviewed articles in professional journals such as *British Journal of Psychiatry, Journal of Neuroscience, Nature Neuroscience, The Lancet, Journal of Biological Chemistry, Neuroscience, Genomics,* and *Molecular and Cellular Neuroscience.* Dr. Pangalos completed his undergraduate studies in Biochemistry from the Imperial College of Science and Technology and earned a Ph.D. in Neurochemistry from the Institute of Neurology, both at the University of London.

Steven Marc Paul, M.D., is executive vice president of science and technology and president of the Lilly Research Laboratories of Eli Lilly and Company. Dr. Paul joined Lilly in April 1993, initially as a vice president of LRL responsible for CNS Discovery and Decision Phase Medical Research. In 1996, Dr. Paul was appointed vice president (and in 1998 group vice president) of Therapeutic Area Discovery Research and Clinical Investigation. In this position his responsibilities included all therapeutic area discovery research, medicinal chemistry, toxicology/drug disposition, and decision phase (Phase I/II) medical research. He and his leadership team were responsible for meeting the pipeline performance objectives of LRL and improving R&D productivity, especially in discovery and the early phases of clinical development. In 2003, Dr. Paul was named executive vice president of the Company and president of LRL with responsibility for all R&D at Lilly. In 2005, Dr. Paul was named Chief Scientific Officer of the Year as one of the Annual Pharmaceutical Achievement Awards. Prior to assuming his position at Lilly, Dr. Paul served as scientific director of NIMH. He received his B.A., *magna cum laude*, in Biology and Psychology from Tulane University in 1972. He received his M.S. in Anatomy (Neuroanatomy) and his M.D., both in 1975, from the Tulane University School of Medicine. Following an internship in neurology at Charity Hospital in New Orleans, he served as a resident in psychiatry and as an instructor in the Department of Psychiatry at the University of Chicago, Pritzker School of Medicine. In 1976, he was awarded a research fellowship in the Pharmacology Research Associate Training Program of the National Institute of General Medical Science to work with Nobel Laureate Dr. Julius Axelrod in the Laboratory of Clinical Science, IRP, of the NIMH. In June 1978, he became a clinical associate in the Clinical Psychobiology Branch of NIMH. In 1982, Dr. Paul was appointed chief of the Clinical Neuroscience Branch as well as chief of the Section on Preclinical Studies, IRP, NIMH. Dr. Paul also served as medical director in the Commissioned Corps of the Public Health Service, and maintained a private practice in psychiatry and psychopharmacology. He is board certified by the American Board of Psychiatry and Neurology and has been elected a Fellow in the American College of Neuropsychopharmacology (ACNP), served on the ACNP Council, and was elected president of the ACNP (1999). He also serves on the executive board of PhRMA's Science and Regulatory Committee and is incoming chairperson. Dr. Paul served as a member of the National Advisory General Medical Sciences Council, NIH (1996–1999), and was appointed by the Secretary of Health and Human Services to serve as a member of the Advisory Committee to the director of NIH (2001–2006).

William Z. Potter, M.D., Ph.D., see Planning Committee bio.

Scott A. Reines, M.D., Ph.D., currently is senior scientist, Foundation for NIH, and is involved in various consulting activities related to the pharmaceutical industry. Dr. Reines retired in September 2008 from Johnson & Johnson Pharmaceutical R&D, where he was senior vice president for CNS, Pain, and Translational Medicine. His department included Clinical Pharmacology/Pharmacokinetics and Pharmacogenomics, as well as the CNS, Pain, and Mature Products Therapeutic Areas. Dr. Reines received his undergraduate degree *magna cum laude* in Chemistry from Cornell University, and his Ph.D. in bio/organic chemistry from Columbia University. He was awarded an M.D. from Albert Einstein College of Medicine, and completed his psychiatric residency at Montefiore Hospital in New York. Following his training he joined the Merck Research Laboratories, where he rose to the rank of vice president, clinical research with responsibilities for Psychopharmacology, Neuropharmacology, Gastroenterology, and Ophthalmology. While at Merck he was responsible for the development of numerous medically important drugs, including the substance P antagonist Emend (aprepitant) for prevention of chemotherapy-induced nausea and vomiting, Maxalt for treatment of migraine headache, Sinemet CR for Parkinson's disease, as well as the antiglaucoma drugs Trusopt and Cosopt. During his 5 years at J&J, Dr. Reines had responsibility for the approvals of Invega for schizophrenia, Reminyl Extended Release for Alzheimer's disease, Risperdal Consta for schizophrenia and bipolar disorder, Risperdal for autism, and Topamax for migraine headache and monotherapy in epilepsy. His research groups also led the development programs for the analgesic drug tapentadol, the long-acting antipsychotic agent paliperidone palmitate, and other potential treatments for various CNS disorders. Over the course of his career, Dr. Reines has published in numerous scientific journals, including *Science, JAMA, New England Journal of Medicine,* and the ACNP journal *Neuropsychopharmacology.* He and Dr. Huda Akil recently served as the first co-chairs of the Neuroscience Steering Committee of the Foundation for the NIH Biomarkers Consortium. Prior to that, he served for 5 years as a member of the National Drug Abuse Advisory Council and its Bioethics Task Force.

Paul A. Sieving, M.D., Ph.D., see Planning Committee bio.

Rae Silver, Ph.D., see Planning Committee bio.

William H. Thies, Ph.D., is vice president for medical and scientific relations at the Alzheimer's Association, where he oversees the world's largest private, nonprofit Alzheimer's disease research grants program. Under his direction, the organization's annual grant budget has doubled, and the program has designated special focus areas targeting the relationship between cardiovascular risk factors and Alzheimer's disease, caregiving and care systems, and research involving diverse populations. He played a key role in launching *Alzheimer's & Dementia: The Journal of the Alzheimer's Association*, and in establishing the Research Roundtable, a consortium of senior scientists from industry, academia, and government who convene regularly to explore common barriers to drug discovery. In previous work at the American Heart Association (AHA) from 1988 to 1998, Dr. Thies formed a new stroke division that recently became the American Stroke Association. He also built the Emergency Cardiac Care Program, a continuing medical education program that trains more than 3 million professionals annually. He has worked with the NINDS to form the Brain Attack Coalition. Prior to joining the AHA, he held faculty positions at Indiana University in Bloomington and the University of Pittsburgh. Dr. Thies earned a B.A. in Biology from Lake Forest College in Illinois and a Ph.D. in Pharmacology from the University of Pittsburgh School of Medicine.

Roy E. Twyman, M.D., see Planning Committee bio.

Nora D. Volkow, M.D., see Speaker bio.

Frank Yocca, Ph.D., is currently vice president and head of CNS and pain drug discovery for AstraZeneca in Wilmington, DE. His research focus is on new treatments for psychiatric diseases. Dr. Yocca received his Ph.D. in Pharmacology from St. John's University in New York City. His work focused on the effect of antidepressants on circadian rhythms. Subsequently he was a postdoctoral fellow at Mt. Sinai Department of Pharmacology. Prior to joining AstraZeneca, Dr. Yocca was executive director at the BristolMyersSquibb Pharmaceutical Research Institute. Dr. Yocca originally joined the Bristol Myers Company in 1984 as a postdoctoral fellow in CNS research. Using techniques he learned from his academic postdoctoral position, he helped to elucidate the mechanism of action of the anxiolytic drug Buspar. He then joined Bristol Myers and made significant advances in understanding the physiological role of the 5-HT1A receptor and its role in psychiatric disease states. During the 21 years Dr. Yocca spent with Bristol-Myers and then BristolMyersSquibb, he supported a number of psychiatric discovery programs, helping to discover and develop the antidepressant drug Serzone. Throughout his tenure, Dr. Yocca continued to work in the field of serotonin and advanced a number of agents to clinical trials, including several antimigraine agents (avitriptan) as well as antipsychotics and anxiolytics. In the latter stages of his career at BristolMyersSquibb, Dr. Yocca became involved in externalization and development. He contributed to the in-licensing and development of the antipsychotic agent Abilify. Additionally, Dr. Yocca was part of the externalization team that in-licensed to Bristol-MyersSquibb the recently approved antidepressant agent Emsam, the first antidepressant to be administered through a patch. In development, he was early development project leader for CRF antagonists and was involved in Phase IV clinical trials with Abilify. Dr. Yocca is a member of numerous scientific societies, including SFN and ACNP.

Christian G. Zimmerman, M.D., F.A.C.S., M.B.A., is chair and founder of the Idaho Neurological Institute (INI); adjunct professor of psychology at Boise State University; and past chief executive officer of Neuroscience Associates. He has also served as a board member for the Idaho State Board of Health and Welfare. Dr. Zimmerman established the INI research facility to focus on nervous system injury, repair, and neuroplasticity. He leads its various interdisciplinary research teams and is coprofessor for biology and cognitive neuroscience research students trained at the facility. Research projects include a 20-year longitudinal study of traumatic brain injury, investigations of spinal injury, stroke, aneurysms, arterial thrombolytic therapy intervention, neuropathology, CNS tumors, sleep disorders, deep-brain stimulation, movement disorders, and five TATRC telemedicine grants. In his role as INI chair, he has facilitated numerous symposia and workshops to provide educational opportunities for medical professionals and the public. Dr. Zimmerman is a diplomate of the American Board of Neurological Surgery and Pain Management and a Fellow of the American College of Surgeons and Physician Executives. He received his M.B.A. from Auburn University.

STAFF

Bruce M. Altevogt, Ph.D., is a senior program officer on the Board on Health Sciences Policy at the IOM. His primary interests focus on policy issues related to basic research, and preparedness for catastrophic events. He received his doctorate from Harvard University's Program in Neuroscience. Following more than 10 years of research, Dr. Altevogt joined The National Academies as a science and technology policy fellow with the Christine Mirzayan Science & Technology Policy Graduate Fellowship Program. Since joining the Board on Health Sciences Policy, he has been a program officer on multiple IOM studies, including *Sleep Disorders and Sleep Deprivation: An Unmet Public Health Problem*, The National Academies' *Guidelines for Human Embryonic Stem Cell Research: 2007 Amendments*, and *Research Priorities in Emergency Preparedness and Response for Public Health Systems*. He is currently serving as director of the Forum on Medical and Public Health Preparedness for Catastrophic Events, the Neuroscience and Nervous System Disorders Forum, and as a co-study director on the National Academy of Sciences Human Embryonic Stem Cells Research Advisory Committee. He received his B.A. from the University of Virginia in Charlottesville, where he majored in Biology and minored in South Asian Studies.

Andrew M. Pope, Ph.D., is the director of the Board on Health Sciences Policy at the IOM. With a Ph.D. in Physiology and Biochemistry, his primary interests are in science policy, biomedical ethics, and the environmental and occupational influences on human health. During his tenure at The National Academies and since 1989 at the IOM, Dr. Pope has directed numerous studies on topics that range from injury control, disability prevention, and biologic markers to the protection of human subjects of research, NIH priority-setting processes, organ procurement and transplantation policy, and the role of science and technology in countering terrorism. Dr. Pope is the recipient of the National Academy of Sciences President's Special Achievement Award and the IOM's Cecil Award.

Sarah L. Hanson is a senior program associate for the Board on Health Sciences Policy at the IOM. Ms. Hanson previously worked for the Committee on Sleep Medicine and Research. She is currently the senior program associate for the Forum on Neuroscience and Nervous System Disorders. Prior to joining the IOM, she served as research and program assistant at the National Research Center for Women & Families. Ms. Hanson has a B.A. from the University of Kansas, with a double major in Political Science and International Studies. She recently completed a post-baccalaureate pre-med program at the University of Maryland and hopes to attend medical school.

Lora K. Taylor is a senior project assistant for the Board on Health Sciences Policy at the IOM. She has 15 years of experience working in The National Academies. Prior to joining the IOM, she served as the administrative associate for the Report Review Committee and the Division on Life Sciences' Ocean Studies Board. Ms. Taylor has a B.A. from Georgetown University with a double major in Psychology and Fine Arts.

Dionna Ali served as an Anderson intern. For the past 2 years, she has worked with Daniel Talmage of the Air Force Studies Board within the Division on Engineering and Physical Sciences. This past summer she also interned at the Kaiser Family Foundation, working with Sean Wieland and Hillary Carrere of the Technology Working Group. At Kaiser, she helped arrange conferences and produce webcasts about health care policy, Medicaid/Medicare, and HIV/AIDS. Ms. Ali is a junior at the University of Virginia, where she majors in anthropology and is preparing for medical school. She aspires to become a neurologist or a psychiatrist.